Failure Mapping

A New And Powerful Tool For Improving Reliability And Maintenance

Failure Mapping

A New And Powerful Tool For
Improving Reliability And Maintenance

Daniel T. Daley, P.E.,CMRP

Industrial Press Inc.
New York

Library of Congress Cataloging-in-Publication Data

Daley, Daniel T.
  Failure mapping : a new and powerful tool for reliability and maintenance
/ Daniel Daley.
      p. cm.
  Includes bibliographical references and index.
  ISBN 978-0-8311-3386-3 (hardcover)
  1. System failures (Engineering) 2. Reliability (Engineering) I. Title.
  TA169.5.D35 2009
  620'.00452--dc22
                                   2008052940

Industrial Press, Inc.
989 Avenue of the Americas
New York, NY 10018

Acquisitions Editor: John Carleo
Copy Editor: Bob Green
Interior Text and Cover Design: Janet Romano

10   9   8   7   6   5   4   3   2   1

Dedication

To my mother and father, who helped create the
map I followed through my life.

Table of Contents

Introduction

Failure Mapping is a system that recognizes patterns of events or conditions that have a direct relationship with subsequent failures. The ability to describe the meaningful elements of those patterns with high accuracy and then to recognize the relationship between them and a failure depends on the ability to capture simple unadorned facts. This book describes a process that fills those needs more completely than other approaches by detailed examination of events, a process that is known as Failure Mapping.

Not too long ago, a company had, what was then viewed, as two competing reliability processes. One was Root Cause Analysis (RCA) and the other was Reliability Centered Maintenance (RCM). Rather than being in competition in terms of value or usefulness, these processes were viewed as competing for the same limited resources. The company had a number of reliability options and only a limited pool of individuals available to pursue the solutions.

The people who supported RCA held the position that it would provide a sure return on their investment in a short period of time. The people who supported RCM argued that the opportunity for improvement was much broader than could be effectively addressed by RCA and that RCM would provide a much more comprehensive solution. It turned out that both sides and neither side won. The RCA program that was implemented held that only a small number of "true disciples" could act as RCA facilitators, so rather than being an "inclusive" process, it turned out to be an "exclusive" process. The RCM program was a healthier, inclusive process, but it was starved for the resources needed to make it a real success.

Viewed from the perspective of someone in the reliability business, the dichotomy described above is silly and wasteful of resources and time. Both RCA and RCM are important tools and are critical elements of any comprehensive reliability program. On the other hand, we can use the competition to provide a useful starting point in analyzing both RCA and RCM for their strengths and weaknesses. The argument described above is one that is not uncommon during the early days in the development of a comprehensive reliability program for any company. The reason for the argument is not so much about people being perfectly happy with their proposed solution as it is about people being dissatisfied with the alternative. If there was a perfect solution, they all believe they would choose it.

STRENGTHS AND WEAKNESSES OF RCA AND RCM.

RCA is one of the most basic and reliable tools for identifying the true cause of any failure. Beyond that, RCA is probably the single best tool for teaching non-engineering personnel the rigor and discipline of the scientific method. This issue is critical for companies that depend on complex capital-intensive assets for their success, but where most employees do not have engineering backgrounds.

RCA is a relatively quick and sure technique for finding the cause and resolving troublesome problems. Drawbacks are that RCA can be resource intensive and the results can address only one specific issue. In some applications it is possible to extend what is learned from one study to a number of problems that share a common cause, but that is more a matter of good fortune than design.

The greatest value stemming from RCA, that has general applicability, is broadening the ability of the organization to recognize patterns that contribute to failures and the relationships between those patterns and specific ensuing failures. In a work environment that is not more than two generations from blaming failures on "gremlins," and one generation from having the attitude that "bad things happen." It is critical that people clearly understand that failures are the result of defects and defects result from specific preventable causes.

The biggest limitation of RCA is that it is very much after-the-fact. The solutions are identified long after the problem has occurred and the loss has been sustained. There is a saying that, "Once the cat is out of the bag, you need to think about dealing with a cat and not a bag". RCA is very much a technique that deals with cats and not bags.

The RCM process is very comprehensive and can be used to thoroughly address all devices that experience failures in any system. RCM is highly structured and can be tightly integrated with a variety of data sources. As with RCA, RCM can be helpful in teaching individuals who do not have technical backgrounds the importance of applying the scientific method when  attempting to observe patterns that create defects, and how to assess the relationships between those patterns and the failures that result.

The main benefit of RCM is that it produces the most effective program of predictive and preventive maintenance, and works to harvest all the available inherent reliability of the system being analyzed.

On the other hand, RCM has several shortcomings including:

1. It is time-consuming and resource intensive.
2. Without introducing some enhancements, the resulting reliability is limited to the inherent reliability of the system being studied.
3. As with RCA, it is another process that deals with cats rather than bags. Its results occur long after the fact.

With both RCA and RCM, there is the same recurring theme: dealing with cats and bags. Both these programs go into action after an untoward event has occurred and the event has usually happened some time in the distant past. Both techniques are good and very helpful, but it would be better to have a process that is capable of moving the line-of-scrimmage from after-the-fact to real-time and to be able to act before the failure has an opportunity to occur.

So, looking at the good and missing features of both RCA and RCM, it would be helpful to describe the features of a process that fills all those needs. The process needs to be:

1. Inclusive – it needs to involve the widest population for understanding and applying the scientific method in dealing with defects that lead to failures.
2. Rigorous – it needs to produce results that will withstand scrutiny.
3. Disciplined – it needs to be based on a well-defined process that participants can follow and that can be audited when anticipated results are not being achieved.
4. It needs to be done quickly, immediately after the failure; at the same time that the defect is forming, or while the failure mechanism is at work.
5. It needs to be based on an accepted structure that participants can apply in a manner consistent with the accepted science of reliability.
6. It needs to be comprehensive – it needs to be part of a process that touches all systems and equipment and not just a select few.

The process described by the requirements listed above is otherwise termed Failure Mapping. Failure Mapping is a technique that applies historical experience to create maps showing how failures occur, then it uses those maps to observe and intervene in the cycle of on-going deterioration and failure patterns to prevent failures whenever possible and to minimize their impact when they cannot be avoided.

The objectives of this book are to:

- Describe the overall concept of Failure Mapping.
- Understand where some elements of Failure Mapping are already being used in systems used by an organization.
- Identify how to perform an assessment of the value that more rigorous Failure Mapping will provide to the company.
- Identify how a firm might go about implementing a Failure Mapping process.

As a professional who has been involved in the reliability business for quite some time, the author believes that the process of Failure Mapping is the most exciting new tool to come down the pike for quite some time. As discussed in the paragraphs above, although not a replacement for either RCA or RCM, Failure Mapping offers many of the characteristics of both those processes, and a solution to cure their shortcomings.

In this book, the author has taken the liberty of creating a virtual organization that shares among its members a culture that is based, in part, on an understanding of the value of Failure Mapping. Rather than having a vague sense that gremlins cause failures, or that bad things are bound to happen, this culture understands that there is a definite sequence of causes and effects that ultimately lead to each and every failure. In the culture that is used as a basis for comparison, the members of the organization are willing and able to use their understanding of the sequence of events leading to failures to intervene before failures occur, or to respond in a manner that absolutely minimizes their impact. Although the author has experienced all the elements that are described in this book in one organization or another, none of those organizations has succeeded in putting all the pieces together into a comprehensive system that will apply all the concepts to the fullest extent possible. The accomplishment of that objective would xiv

be the result of the Failure Mapping environment described herein.

The author has also taken the liberty of tightening up some concepts that he feels have long been too loosely defined, to the detriment of the structures and discipline needed to accomplish the organizational and cultural objectives described above. For instance, the concepts of diagnostics and troubleshooting are two functions that are frequently only loosely defined and, as a result they are frequently accomplished by a range of individuals in a variety of roles with little structure and in a hit-or-miss fashion. This book focuses on those two roles as keys to being able to quickly respond to failures and to apply the proper solution(s). Other examples of poor definitions are those for terms like Malfunction Report and Failure Mode. In this context, those two terms represent the specific starting points (Affected Function – Current Behavior), and ending point (Defective Component – Component Condition), for the portion of the Failure Map that is used to support immediate response to failures.

As with several other books in this series, the author has written the current book in a manner that makes it a useful tool for the reliability professional and other individuals who depend on reliability for their success, and the successful use of assets and organizations that they manage. The book is short enough to be used by busy people, but it provides sufficient detail to completely describe the subject being addressed.

For those readers who are not familiar with the concept of Failure Mapping and how it is applied, a brief summary is included in the appendix. It is hoped that readers will enjoy this book and find its contents useful to them and their organizations.

Chapter 1
Patterns and Relationships

"Nature uses only the longest threads to weave her patterns, so that each small piece of her fabric reveals the organization of the entire tapestry." **Richard P. Feynman**

Every day, the TV business news features two or three individuals who purportedly are able to tell the audience why the stock market went down or up that day. Although some of these individuals are regulars, or have been on TV several times, many are individuals who are on TV only once and who have earned their fifteen minutes of fame by having been in the stock advice business for a long time. The opportunity to be on TV is probably a plum for them because it provides them with some free advertising; maybe some name recognition, and ultimately a few new customers. The belief is that, if these individuals can explain what just happened to the market, they likely as not understand it well enough to recognize impending events before they occur and their customers will prosper from that foresight.

If you watch these shows closely, you will soon realize that these so-called experts simply identify some event or combination of events that form a reasonable hypothesis for the cause of the change, and they describe their hypothesis in a very compelling manner. In some instances, the event that is identified as the cause of the change reverses itself the very next day, but the general stock market does not reverse itself. In other words, there is no well-defined relationship between the recognized pattern and the following event.

Ask yourself: if you were able to identify a variety of patterns that had direct relationships with specific changes in the stock market, how would you use the information? Would you be on TV trying to con-vince the audience how smart you are? For me the answer to the first question is that I would use that knowledge to make a ton of money for myself. The answer to the second question is that, rather than being in a TV studio, I would be on a yacht, voyaging around the world and phoning in a few critical choices to my brokers to keep my pile of money growing. Obviously, the more people I let in on my secret, the more I would dilute my own results.

In other words, I don't believe these experts really know the causes and effects. In the financial world, there are no absolute relationships. There are recognizable patterns, but the element of human emotion tends to weaken the tie between any recognizable pattern and a specific event.

One might ask: if the financial example does not meet the gold standard as an acceptable form of "patterns and relation-ships," then what does? There are myriad examples of instances where a pattern of circumstances or events is always followed by a specific event and the relationship is so certain that it can be viewed as cause-and-effect. Several examples are:

• Corrosion of a pressure-retaining device resulting in a failure.
• Repetitive fatigue cycles leading to failure.
• Exposure to loads beyond the design limits, leading to failure.

In an environment in which all failures are not studied in a scientific manner, it is possible that patterns are concocted and relationships are imagined. A good comparison is with gremlins causing problems, or angels providing assistance when needed. Although those are extreme examples, many people do get into the habit of relating failures to circumstances that may have little if any bearing on the failure.

In his book "How Do We Know What Isn't So" Thomas Gilovich provides a number of examples of everyday life situations in which people see patterns leading them to beliefs that relationships exist that are totally unfounded. An interesting example has to do with previously-barren couples who adopt a child and then the wife becomes pregnant. According to the non-scientific system of beliefs, the fact they have adopted a child allows them to "relax," and then for nature to take its course, leading to a pregnancy.

In this situation, so long as both members of the couple are healthy and capable of conceiving, they will either have a child or they will not. The statistical likelihood of conceiving remains the same independent of the adoption. If the adoptive parents later conceive, that fact tends to support the biased belief system and adds one point in favor of the notion. If they do not conceive, the situation merely confirms the logic behind the adoption. Thus, no points are scored against the unscientific notion.

In an industrial plant, it is not uncommon to look for a relationship between poor performance and the individuals who are on duty when a problem occurs. With heavy mobile equipment like locomotives, it is not uncommon to look for a relationship between weather and failure frequency. Although those factors may occasionally appear in a pattern leading up to a failure, more frequently they do not.

With poor operators on duty, it is most likely that a number of other elements making up a pattern leading to a failure already exist, and the presence of the poor operator is the final element. One or more of the poor operator's marginal practices evidently resulted in conditions that were no longer tolerable by the already weakened systems.

Most locomotives are built to make it through any kind of weather. Extremely hot, cold, or wet weather tends to identify the units that have existing weaknesses. The failures occur because of the weaknesses, not the weather. Weather will happen. Although it is not "ordinary," it is expected.

In a business setting, there are two important characteristics that lead to what one of my professors used to call "stinkin'-thinkin'." as exemplified by the following circumstances:

1. The rigor of the scientific method has not been applied to under standing any of the perceived "patterns and relationships".

2. There is no organized system that forces rigor to be applied to all circumstances and events in a disciplined manner.

As a result, life goes on and people remain insistent upon the validity of their beliefs, even when there is no real data to support them. In a home or personal setting, these assumptions lead to problems associated with normal human biases such as:

- Racism
- Ethnic differences
- Sexual biases
- Etc.

Although those issues are troublesome, it is well beyond the objectives of this book to address them. Conversely, those same problems exist in asset-based business operations, and it is the objective of this book to address those problems.

Most, if not all, things fail as the result of a consistent pattern of events. The elements of that pattern are different for each and every failure, but the pattern is consistent. If the pattern is understood, and the events in the pattern are treated in a consistent manner, a starting point exists for quickly dealing with the failures and even eliminating them before they occur.

The starting point is to begin ridding people and organizations of non-fact-based, unrealistic belief systems. This exercise probably sounds easier than it is. People feel comfortable with their beliefs, and giving up their biases is tough. Often, personal identity and esteem is built upon those biases and asking that they be relinquished leaves the person in uncertain territory.

Suppose I am an operator and I have a belief that the systems I operate are inherently unreliable. I believe that these systems fail on occasions because of unreliability and are in no way due to my poor operation. Based on that scenario I can continue to maintain a belief system that supports my self esteem as a capable and knowledgeable operator.

Now you come along and tell me that something I am doing is causing deterioration and, ultimately, failures. My belief system tells me that good operators do not cause failures. It is important to me to view myself and be viewed by others as a good operator. The two views are inconsistent. I cannot both believe you and continue to maintain the belief system I have held for my entire career. This contradiction causes what is called a "personal crisis". It is a situation in which I am learning for the first time what the world around me has known for quite some time.

As stated above, this dilemma is the starting point. All members of an organization must have this kind of epiphany, and they must become so hardened to the painful truth revealed in dealing with the facts of each and every situation that these facts are no longer painful.

It is far easier to deal with information that you only half believe and that is probably why the TV commentators mentioned at the beginning of the chapter continue to be employed. Can you imagine how the world would be if everyone in the public media described only exact facts? First, they would be confined to saying very little because the portion of what they say that is certain fact is quite small. Second, the public reaction to their revelations would be devastating. The TV commentator would say that a recession or depression was coming and everyone would go into hiding. Our intuition that most people exaggerate their knowledge is helpful in situations where a significant portion of what is being said is for entertainment value.

The problem is that this TV mentality tends to intrude into our work and business lives. We allow people to exaggerate in situations where facts and only facts should be the standard. But to administer this standard of facts it will be necessary for people to say only what they know. To ensure that individuals say only what they know we must ask for information in a different manner. Our manner of asking must restrict the answer to the range of possible facts.

Think about it, when a person in power asks a question, there is an implied expectation that an answer can be provided. Many of the structured systems used by our organizations contain a further implication that the questions are being asked by the individuals in power. When something breaks, a failure reporting system asks a person to describe what went wrong. The conditions created by the system compel the person to provide a response. Not knowing what form of response is expected, the respondent provides a colorful description that is primarily intended to distance him from any possible blame. As a result, a lot of useless information is provided.

Added color and added distancing in the description creates a starting point that cannot be reconciled with other data to describe a starting pattern for mapping failures.

The same situation often exists at the endpoint of a failure that has been repaired. The greatest desire on the part of both the individual and the organization is to tackle the next problem rather than taking the time to document the previous one. As a result, the endpoint of the failure map is left unpopulated or only poorly populated, and mapping is impossible.

Again, this sequence of events creates the need for yet another epiphany for individuals performing the repair, documenting findings, and searching for cause. Complete organizations as well as each and every member need to be committed to collecting the information needed to create failure maps.

In the distant past, there was a detective series on television titled Dragnet. As the show's leading character, Sergeant Joe Friday was remembered for the guidance he always provided when asking questions. Joe always said, "Just the facts" to emphasize how important it was for people to only describe what they knew for certain.

As stated above, Failure Mapping is a system that recognizes patterns of events or conditions that have a direct relationship with subsequent failures. Our ability to accurately describe the meaningful elements of those patterns and then to recognize the relationship between them and a failure depends on our ability to capture simple, unadorned facts.

Chapter 2
The Path to Failure

"Failure should be our teacher, not our undertaker."
Denis Waitley

Over the course of the author's career there have probably been thousands of times that someone has said, "As soon as you explained the problem, I could have told you what was wrong." In those instances, the speaker was describing a situation in which a clear understanding of a set of symptoms or behaviors immediately opened the door to an understanding of the Failure Mode. In effect, this individual's personal experience had allowed him to create a mental failure map that directly associated a specific form of malfunction to a specific, most-likely failure mode.

Individuals involved in the business of maintenance or reliability eventually begin to recognize patterns in their activities. In some instances, the patterns become apparent because the same equipment fails over and over again. Those situations are fairly obvious. On the other hand, there are also many patterns that require greater insight or more effort to identify.

If your organization is one that regularly performs Root Cause Analysis, you may have tracked a number of failures back to a single systemic cause that created the circumstances leading to a variety of problems. Here, hard work and dedicated effort has highlighted the commonality of otherwise transparent causes. At other times, it may be necessary to assign dedicated resources to produce those findings. The old saying often applies: "It is difficult to remember that your job is to drain the swamp, when you are up to your neck in alligators". Saying it differently, it is difficult to step back calmly and recognize all the patterns and relationships that may exist when your primary work is to keep the assets operating.

Despite the fact that they are difficult to recognize at times, it is important to keep in mind that all failures are the result of a series of steps and that those steps can be tracked and characterized by a consistent set of descriptors.

The fact that those steps follow a similar path is a very good thing indeed because it can provide the ability to create a consistent way to characterize the steps that follow a path to failure. By characterizing the steps in a consistent manner we can catalog and monitor the relative frequency of repetitive events. Understanding those patterns will provide us with the opportunity to get ahead of the game and prevent failures before they occur.

It is not uncommon for different organizations, and even different groups in the same organization, to use different ways to describe failure patterns. The result is limited success in collecting and interpreting data.

This chapter is intended to provide the reader with a useful way to view the path of events that ultimately leads to failure. By using this discipline it is possible to collect and characterize data and then manipulate it to produce useful tools for preventing failures and for managing failures when they occur.

Systemic Cause
The beginning of any failure can usually be found in a systemic of latent cause. A systemic cause is a gap in an organization,

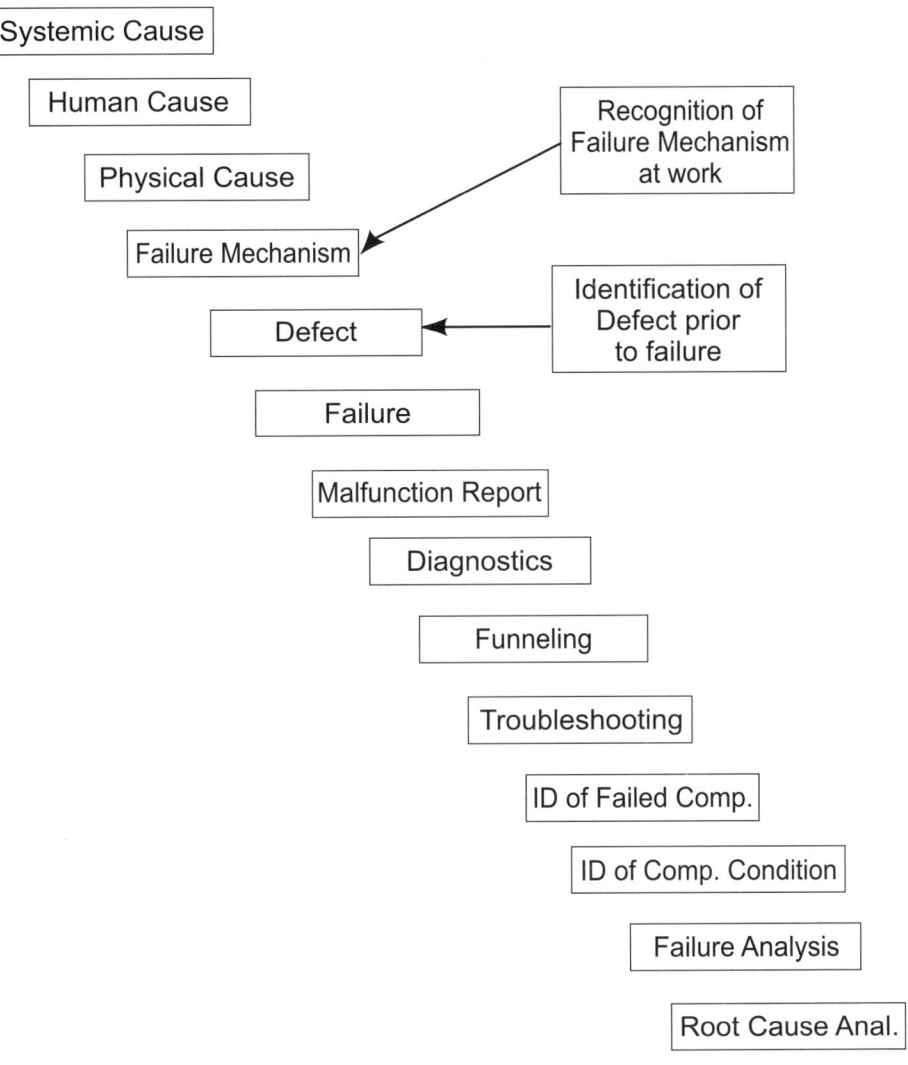

procedures, or processes that creates a trap for someone to step into or for a failure to happen. In some examples, the gap is small so circumstances have to be just right for a failure to occur. Most wide gaps are fairly apparent and are soon closed.

Examples of systemic causes include situations where procedures allow individuals to pass jobs from one person to another without adequate communications. Another example is a situation that depends on the knowledge or skills of the person(s) performing the work when some portion of the individuals who may be assigned to the work are not adequately qualified.

Human Cause

The next step is the combination of circumstances that cause a specific individual or group of individuals to step into the trap and become the human cause of the failure. This step can be an act or an omission, but it connects a person to the failure.

An example of a human cause is when two individuals pass a single job from one to the other without adequate communications of status. The individuals may be following all the written procedures but writers of these procedures sometimes assume that everyone will naturally "go the extra mile" or take extra steps to cover for shortcomings in the organization. When someone is tired or simply forgets to cover all the details, they fall into the trap and become the human cause.

Physical Cause

The next step is the physical cause. Often, the search for cause, or the root cause analysis, stops when the physical cause is found. Drilling down below the physical cause to identify the human cause and the systemic cause frequently becomes messy. Physical systems do not suffer from "blame," but individuals and the managers who are responsible for systems do.

Continuing the example described above, the physical cause can be anything from a support that is left off, to a gasket left uninstalled, to an inspection or quality control step that is missed. In any event, the physical cause is a problem in the product that allows "nature" to act in a "natural" but "undesired" manner.

Failure Mechanism

The next step on the way to a failure is the Failure

Mechanism. The event that establishes the physical cause starts a Failure Mechanism working and if that Failure Mechanism is not arrested, it will ultimately produce a defect. Despite the fact that there is an infinite number of Failure Modes, there is a relatively small number of Failure Mechanisms. This aspect becomes important later in this discussion.

A good aspect of any Failure Mechanism is that there typically are tell-tale signs when the Failure Mechanism is at work. Corrosion leaves rust. Fatigue will engender unusual movement from vibration, or displacement from misalignment. Whatever the sign, it is possible to observe it and intervene before the Failure Mechanism proceeds to the point that it has created a Defect.

Another positive aspect of most Failure Mechanisms is that they typically take time to work and to produce the amount of deterioration required to result in a Defect. For instance, in a piping system, a small amount of corrosion will not produce a failure. The corrosion needs to deteriorate the pipe wall to the point that it is less than the minimum wall thickness needed to retain the maximum operating pressure.

Defects

After a Failure Mechanism has been at work for an appropriate amount of time, it will produce a defect. The defect mechanism begins the statistical process that determines when a failure will occur. As with the Failure Mechanism, many defects do not immediately result in a failure. For instance a small crack might not immediately result in a leak. The presence of the crack may need to be combined with a situation when unusually high (but still expected) pressure levels occur. Like the Failure Mechanism, the defect is an anomaly that can be detected, and as such it provides another opportunity for steps toward prevention.

Failures

With the presence of a defect and a certain set of circumstances, a failure will occur. In the reliability business, a failure is the situation in which one (or more) of the critical functions is lost.

Failure is the next step in the path and is separate and distinct from both the formation of a defect (which precedes it) and the issuance of the Malfunction Report (which follows).

It is important to separate the formation of the defect from the failure, and further separate the failure from the issuance of the Malfunction Report. As will be discussed, these three non-simultaneous events (defect, failure, malfunction report), are separated by time and by the statistics of nature. The effectiveness of any reliability program and the bottom-line reliability and availability of any assets can depend on how well the opportunities presented by detection of these events are captured.

Malfunction Reports

After the failure occurs, there is some span of time before it is recognized and reported. Although failures can be reported in a number of ways, the most useful information describes the function that has been lost and the behavior that is being exhibited. The important thing is to report all that is known for certain, but no more.

It is not uncommon for there to be very little structure or discipline associated with the manner in which a failure is reported. Failure reports often use colorful descriptions that highlight individuals' concerns about surrounding events, rather than describe the lost function and the problematic behavior.

Diagnosis

After the malfunction report has been issued, diagnosis can begin. For purposes of this discussion, it will be assumed that diagnosis is an act that occurs remotely using only information that can be assembled remotely. If there is a dominant failure mode and plenty of experience with the malfunction report (such as function lost and specific behavior) associated with an occurrence of that failure mode, it might be possible to point to the ultimate solution quickly. In other circumstances, the initial diagnostic step will only point to the one or more systems where the failure could have produced the behavior being experienced.

Funneling

Typically, the next step consists of funneling. As in the sequence, Ready – Aim – Fire, there is a rationale behind the sequence Diagnose – Funnel – Troubleshoot. The objective in funneling is to reduce the focus from a major system to a sub-system, or specific equipment item, before beginning to perform the time-consuming and resource-consuming step of troubleshooting. Funneling is a procedure that can depend on added information like computer control system fault codes, and failure records (maps) from earlier experiences. Additionally, there may be specific characteristics of the behaviors that point toward one specific system. For example, a diesel-electric locomotive can exhibit a "loading problem" that results from either an engine failure or a failure of the main alternator. If the engine has been pouring heavy black smoke out of the exhaust at the time of the failure, the apparent symptoms point to an engine problem.

Troubleshooting

The next step is troubleshooting, which is the time- and resource-intensive step that may entail disassembly of major systems to identify the failed component and its condition. False steps in troubleshooting are costly. As a result, more-accurate funneling may lead to greater effectiveness and efficiency.

In the example of the locomotive described above, heavy black smoke at the time of the failure would lead the diagnostician to recommend troubleshooting the engine before anything else. While that step might seem obvious, it is useful to keep the real-life situation in mind. In the real- life situation, the work of the troubleshooter may be far removed from when and where the failure took place. The engine may not be running, so it would be impossible to observe or apply instruments to observe the symptoms of the systems in operation. To a troubleshooter, a dead, cold main alternator looks equally as faulty as a dead, cold engine.

The next two steps are companions.

Identification of Failed Component(s)

The first of the steps is identifying the failed component, and the second is assessing the condition of that component. For purposes of this discussion, that combination will be used to describe the Failure Mode. An unfortunate feature of current-day maintenance is that many technicians are actually only "parts-changers". As a result, these people continue to change parts until things begin to work. Sometimes, as many as two-thirds of the parts changed in this manner are good and were not the cause of the failure. Challenging mechanics to identify not only the defective part but also the aberrant condition, is a way to avoid the parts-changing pitfall.

Identification of Component Condition

The following example is intended to emphasize the importance of separating the act of identifying the failed component from that of identifying the condition of that component. With electronic systems using plug-in cards or-modules, it is not uncommon for apparently-good cards or modules to be replaced, only to have the device begin operating normally. If there is no testing device to verify that the board or module has failed, the cause of the failure is not really known. The component connectors could have become loose or the connectors could have become deteriorated or coated with something introducing resistance. In these conditions, it is better to describe the failure mode as "Device X- changed and operation restored". This report leaves the condition of the device open to future determination when better testing is possible.

The next two steps are on the "path to solution".

Failure Analysis

The first of these steps is Failure Analysis. For purposes of this discussion, Failure Analysis will be the step of identifying the Failure Mechanism. As mentioned earlier, there are only a handful of Failure Mechanisms, so the task is somewhat simplified by knowing the clues and characteristics of each possible Failure Mechanism. For example, the presence of rust will point to corrosion as the Failure Mechanism.

Although some of the other Failure Mechanisms may not have as pronounced or evident clues as does corrosion, an experienced investigator soon gains the ability to identify those signs that lead to the "Path to the Solution". The following are a few examples:

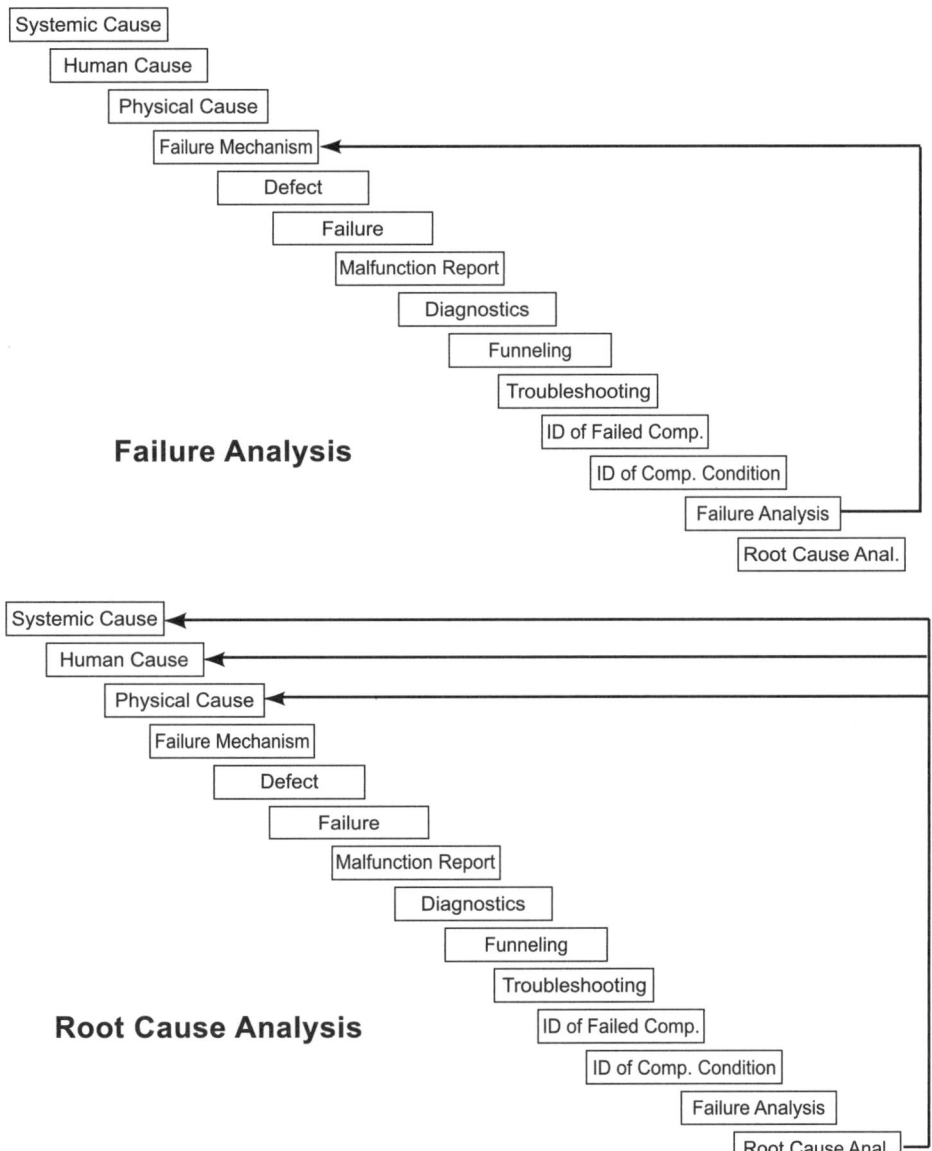

- Scuffed paint where a support was left off.
- Signs of overheating.
- Evidence of leakage

Root Cause Analysis

The second step is Root Cause Analysis, which is the process of identifying the three levels of cause:

- Physical Cause
- Human Cause
- Latent or Systemic Cause

The ability to prevent the same failure from happening again and to prevent all similar events is directly proportional to your organization's effectiveness in finding the root causes of failures.

Both Failure Analysis and Root Cause Analysis depend on having trained and dedicated resources in the organization, who are assigned to perform those steps. It is not realistic to expect that the people performing the repairs will also perform the investigation. Repair people typically get on to the next thing that needs to be repaired, or the next item of "real work" on their list. For individuals who are assigned "urgent" and tangible tasks, tasks with intangible results like finding cause seem unnecessary and unfulfilling.

Example

What follows is a relatively simple example including all the steps described above. Although the example is simple, it is important to realize that the process described is applicable to a wide variety of problems. The approach is most valuable when applied to highly-complex problems and in an environment that has many and varied problems. The presence of those characteristics emphasizes the importance of consistency and the ability to catalog and easily recognize similar situations.

Latent or Systemic Cause

Let's assume that a plant has no specific procedure guiding

the installation of "all" rotating equipment. There are no clear requirements that pumps or compressors be
aligned or balanced, or that suction or
discharge flanges be checked for
"cold-spring," or that bases be
checked for "soft-feet" as part of the
check-out procedure.

Human Cause

Now let's assume that the Project Engineer, Fred Willis,
installs a new low-speed compressor as part of a project. While the
size of the equipment is impressive, it operates at less than 400
r.p.m. so that Fred (as well as all his advisors), believes it is not
necessary to balance the coupling or use anything other than a
very rudimentary technique for aligning the compressor shaft with
the driver shaft.

Physical Cause

In this specific situation (which is based on a real-life experience) the combination of imbalance and misalignment created a
situation that led to fatigue. During the investigation it was found
that when they were installed, the individual coupling halves were
enough out of balance to cause the rotating assembly to roll, without applying any other forces, Also, the alignment was so bad that
the mechanic needed to use a "persuader" to install the coupling
bolts.

Failure Mechanism

Although the compressor speed was
relatively slow, the combination of imbalance
and misalignment was sufficient to produce
stress levels in the compressor shaft that were
above the fatigue limit. Whenever either of the
compressors (primary or spare) was in opera-

tion, it was subject to fatigue cycles at a rate of 400 per minute,
24,000 per hour, or 576,000 per day …. And so on.

(In the real world, where this situation occurred, there were two such compressors, a primary and a spare. Nominally, each machine was operated for half the time. The individuals involved with the machines had a total of twenty-seven years to recognize the problem and take corrective action before the failures took place, but they did not. Both compressors failed in almost exactly the same manner just six months apart.)

Defect

In a situation of the type described, the defect is a crack that forms in the compressor shaft after a specific number of fatigue cycles (revolutions). There are many sources of literature that will provide the interested reader with the proper way to identify the crack initiation point. In the real-life case being mirrored in this example, the shaft had a key-way and some people thought that the crack propagated from the key-way. Detailed analysis showed that it did not.

The formation of the defect described in this example was found to be almost simultaneous with the time of the failure. All the energy needed to drive the defect to failure was present at the very moment the defect formed.

There are many other situations in which the defect forms and then there is a delay until there is sufficient energy to drive the defect to failure. An example is corroded piping. The pipe becomes corroded beyond minimum wall thickness for the maximum  operating pressure, but the system is not operating at the maxi- mum operating pressure at precisely the time the defect forms. This situation can be compared to "nature" throwing dice. When the defect forms, the first dice has settled on a one. After that, nature continues to throw the second die until it comes up one. At that point, the player experiences "snake-eyes" and loses.

Failure

Snake-eyes! In this example, the formation of the defect

and the failure were almost simultaneous. Despite the fact the compressor shaft was large (approximately four inches in diameter), the centrifugal force due to the coupling imbalance was sufficient to cause the crack to propagate in seconds and to fracture the shaft. In other examples, the defect has had to wait for an exceptionally high (but still expected) load to occur to create a failure. In the present example, the odds were small and the failure was nearly instantaneous with formation of the defect.

Malfunction Report

There was very little subtlety in the malfunction report for this example because the compressor failure resulted in a plant outage. A portion of the broken shaft separated from the compressor and the other coupling half and it was readily apparent to the operator that the shaft had broken.

For each instance like this one, there are numerous others where the Failure Mode is obscure. For example, the shaft might have broken in a different manner and remained in the compressor, and continued to turn but did not turn the impeller. If the operator had reported a broken compressor shaft, he would have been saying more than he knew.

When people say more than they know, the information is typically misleading and often does more harm than good.

Diagnostics

Let's assume that the operating data has been stored and that it is possible to review critical conditions leading up to the outage. In the example, the process information prior to the event would have provided little information other than that the pressure in the header leading from the compressor dropped precipitously at the moment of the failure.

If the system had been equipped with vibration sensors, they would have registered a dramatic increase in vibration in the moments before the failure. Also, because this installation was electrically-driven, any instrumentation used for monitoring the power supply would have shown clearly that there was no interruption in the power supply. In fact, if current monitoring for the drive motor had been available to the operator in the control room, he would have been able to see that the motor was still operating albeit at a much-reduced, no-load, current level.

Funneling

This example is not the most descriptive for the value of funneling but it provides some insight. Here were two compressors that failed in almost the same manner. When the first compressor failed, the shaft detached itself and landed fairly close to the compressor so that the operator could see plainly what had 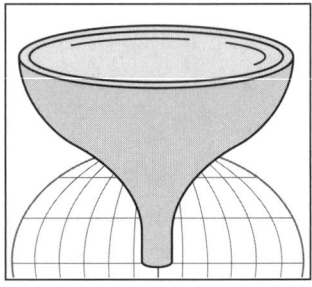 failed. In the second compressor, the shaft broke but remained in place, with the compressor shaft simply spinning in its bearings. The driver shaft was no longer attached to the other half of the shaft, or to the impeller, so the compressor output stopped. Having experienced the other compressor failure only six months earlier, the operator had good reason to believe the failure was due to a similar cause.

To confirm that fact, the operator rotated the portion of the compressor shaft attached to the motor and noted that the opposite portion of the shaft did not turn, proving that the shaft had severed.

Troubleshooting

In the example being described, had the second failure happened first there would have been more intrigue for the troubleshooter. In the second there were no apparent external signs of damage (other than that manual rotation of one shaft had no effect on the other shaft). This installation was made up of heavy equipment and almost every move required special tools and mobile lifting equipment. As a result, had individuals looking for the defective component started dismantling equipment without a clear idea of the cause, a great deal of time and limited resources would have been wasted.

It was important that the troubleshooter be able to recognize a very likely area of failure.

Identifying the Failed Component

Continuing with the compressor example, it was quite simple to identify the failed component. In the first instance, the shaft had separated and was lying some distance from the compressor. In the second instance, both obvious signs and past experience with the companion compressor led to quick identification of the defective component.

This rapid action does not always take place. In most breakdowns the failed component is not a four-inch thick shaft that has been completely severed. The failed component is often some small electronic item or assembly that does not even experience sufficient heating to show signs of overheating. In these events, the troubleshooter must use sophisticated instruments that show "no output with normal inputs" for a complete electronic module or board. In such a failure it may not be possible to identify the exact component that failed (e.g. a resistor, capacitor, soldered connection, etc.). Instead, identification of the Failure Mode must remain at the subsystem or component level rather than the sub-component level.

Identifying the Component Condition

As discussed above, identification is a step that adds another level of difficulty but also adds significant value. In the compressor example, large four-inch shafts that fail in a dramatic manner most often make it into some individuals' collection of memorabilia. As a result, as long as most of the evidence has not been tampered with, it will be possible to describe the condition and perform Failure Analysis for quite some time after the event.

In most situations, easy identification does not happen. Repairable components are returned to the supplier for repair as quickly as possible, to maintain the core pool. Small components and subcomponents with no residual value are discarded, and precious metals are recovered. Normal shop, manufacturing or remanufacturing processes are set up to keep materials flowing in such a manner that evidence is seldom available for very long.

Recognizing these facts, it is important to build steps to aid identification of failed components and their condition into the shop's normal repair process.

Failure Analysis

Although the compressor example provided limited value for several of the steps above, it is an excellent illustration of the value of performing Failure Analysis. As mentioned, two similar compressors failed six months apart.

When the first failure occurred, only a very limited Failure Analysis was performed. There was some passing thoughts that the rather large, roughly-cut key-way created a stress concentration point and contributed to the shaft breakage. To make everyone feel better, a more expensive steel alloy was selected and the shaft was re-manufactured to the same dimensions and installed in the same manner as before.

After the second compressor shaft failed approximately six months later, senior managers became alarmed that there might be a larger, more chronic problem, so additional resources were deployed. Both a Failure Analysis and a Root Cause Analysis were then conducted.

The Failure Analysis identified all the visible signs of a fatigue failure in the shaft. Once fatigue had been identified as the failure mechanism, the analysis continued in a search for the source of the unusual stress level and it identified the imbalance and misalignment conditions.

The study further determined that the new alloy selected to replace the original steel when the first compressor shaft failed was marginally more susceptible to the conditions that caused the failure. No additional steps were taken to balance the coupling or align the shafts or to check for cold-spring or soft-foot, so it was likely that the new, more-expensive, replacement shaft would fail in less time than its predecessor.

As a result of the studies, the replacement shaft on the second compressor was made of the same alloy as the original, the coupling was balanced, the shafts were laser-aligned, and the installation procedure checked for cold-spring and soft-foot. After the second compressor was back in service, these same steps were applied to the first compressor. (Although the twenty-seven year combined life of the two compressors made the return on additional investment for the added work unfavorable, the possibility of collateral damage that might be caused by a failure was viewed as unacceptable. So the improvements were accepted as good business.)

ROOT CAUSE ANALYSIS

As with the Failure Analysis, the Root Cause Analysis for the shaft failure came up with some interesting findings. The fact that the compressors operated at only 400 r.p.m. and that the equipment appeared so robust, created paradigms that set people up for failure, although the term "failure" has to be qualified. After all, the two-compressor system operated without failure for twenty-seven years. In most situations, such a record would have been considered a glowing success. As mentioned earlier, it is not the frequency of failure or the direct impact of failure that gives pause. It is the likelihood of collateral damage and uncontrolled effects that elevate the concern.

The three sections below describe the various levels of root cause starting with the apparent physical cause, drilling down to the human cause then ending with the latent or systemic cause. It is critical to keep in mind that the physical device did not create the reasons for its own demise.

Physical Cause
In our study (as with the real world example), the physical cause would be closely linked to the failure mechanism. For fatigue

to exist and for it to have progressed to the point that it would cause failure of a four-inch diameter steel shaft, there would need to be two factors:

1. There must be sufficient imbalance or misalignment to cause a cyclic load greater than the fatigue limit.

2. The system must be exposed to that loading for an extended period ... sufficient to have involved several billion cycles.

Based on the rotational speed and the hours for which the compressors were in operation, it was apparent that the physical cause had to be something that happened at the time of the original installation (a combined 27 years earlier). Although maintenance work had been done on the compressors several times during their lives, the coupling was never balanced, and the alignment was neither checked nor improved.

As a result, it was apparent that the imbalance and misalignment continuing from the initial installation for the entire life of the unit was the physical cause.

Human Cause
The human cause could have been almost anyone who had been involved in the original construction, installation, or maintenance of the compressors over their entire 27 year life. No one ever identified or corrected the problem. It is possible to be somewhat dismissive in conditions such as this and to say that everyone was at fault, but it is important to be specific enough to support corrective action. If the cause of this failure is to be eliminated, it is necessary to identify each and every person who will need to modify their beliefs and change their behavior. It is important to move from the abstract to the tangible to make real improvements.

Latent or Systemic Cause
The latent or systemic cause is an issue that is an accepted part of culture, and is deeply imbedded in our practices and behaviors. It may or may not be documented in written procedures and training, but that is a secondary issue. What people actually do is more important than what they are supposed to do.

Generally, there is an expectation that people will "do the right thing," but what they actually do is influenced by a wide variety of elements. In the long term, it will be necessary to address all those elements to ensure proper behavior. But just writing a procedure, or telling people what to do, is not adequate. It is necessary to verify that the desired behavior exists.

In this specific example of the compressors, the latent or systemic cause was the widely-accepted belief that low-speed equipment did not require balancing or alignment. This belief was the cause of the coupling not being balanced, and the two shafts being aligned with only the use of a straight edge. This belief permeated the entire organization, even to the extent that there was still resistance to balancing the coupling and performing more sophisticated alignment even after the catastrophic failure of the two shafts.

Convincing those who still felt that balancing and alignment were a waste of time required the team to demonstrate the possible effects of a four-inch diameter hole in the side of a compressor housing at some indeterminate point in time. Although there is a commonly-held belief that it is not necessary to balance and align slow-moving equipment, there is always the need to balance the cost of prevention against the cost of repair and the effects of failure (release of a toxic or hazardous material), must also be considered. Generally speaking, it is easier to establish a consistent requirement for all systems than to expect that the analysis described above will be completed and the proper choices made.

THE PATH TO SOLUTION

Mapping the path to failure in this manner can provide a basis for future solutions. As stated in our objectives, failure mapping can provide a mechanism for improving both maintenance effectiveness and reliability of installations.

Recognizing Failure Mechanisms at Work
    As mentioned earlier, there are only a handful of failure mechanisms. Actually there are two small groups of Failure Mechanisms, one for mechanical systems and one for electrical systems. The advantage of knowing this distinction is that it provides the ability to look for clues that the Failure Mechanisms are at work. If a Failure Mechanism is found to be at work, it may be assumed that it will ultimately create a defect leading to a failure if it is not arrested in time.

    The mechanical Failure Mechanisms are:
- Corrosion
- Erosion
- Fatigue
- Overload

    The electrical Failure Mechanisms are
- Overload
- Supply Transient
- Load Stall

- Electrical Equivalent of Fatigue
    - Persistent loading to levels greater than rated but less than will cause instantaneous breakdown.
- Insulation Breakdown
    - Heat
    - UV
    - Chemical Exposure
- Mechanical Failure
    - Abrasion
    - Loosening

    Regular inspections for tell-tale signs of Failure Mechanisms at work will allow for intervention before a failure-causing defect is formed.

For instance, corrosion leaves corrosion products. Even before a significant amount of corrosion has taken place, leaving products behind, it is possible to identify the elements needed to make a corrosion cell. Start with two dis-similar metals and add a source of liquid to act as the electrolyte and voila, a corrosion cell is created. The presence of dis-similar metals in an assembly is not uncommon. A leaky door seal or a plugged drain will, perhaps unexpectedly, either collect or hold moisture.

Another Failure Mechanism example is fatigue. With pumps and compressors operating at 3600 rpm, the number of fatigue cycles piles up quickly. Poorly-balanced rotating assemblies, or poorly-aligned shafts are not uncommon. These conditions typically evidence themselves in high vibration levels, providing an obvious sign of impending failure.

Finding Defects Before They Cause a Failure

As mentioned earlier, the fact that a defect exists does not mean that a failure will immediately occur. The kind of defect will determine the likelihood of the failure. Sometimes the likelihood of failure remains quite low despite the presence of a defect. At other times, the likelihood is 1:1, meaning that as soon as the defect forms, the failure will occur.

Nature is often kind and provides an opportunity for the defect to be found and remedial action taken before a failure occurs, but vigilance is needed if defects are to be found before they cause failures. A defect in an operating system is somewhat unusual and easy to spot, and there is a belief that any fair-minded defect would show itself as soon as it forms. For instance, a fatigue crack should cause only a small leak when it forms so that people can respond in an organized manner.

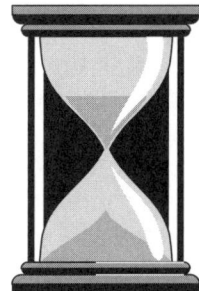

However, some defects are not so apparent and linger for years before they become apparent. As a person who was involved in the Y2K program, the author can report that the team found a number of programming defects that would ultimately have caused failures if not found during testing. For instance, the problem of register or buffer memory overload is one that does not become apparent until the proper string of circumstances begin to align themselves.

In other situations, regular inspection programs or operator-driven reliability programs are helpful in applying enough scrutiny to ensure that defects are at least exposed to several sets of eyes during the time interval after they are formed, but before they can cause a failure.

USING MALFUNCTION REPORTS AND PAST MAPS TO IDENTIFY FAILURE MODES

Apart from finding evidence of Failure Mechanisms at work or defects that have formed, the concept of understanding and tracking failure paths has one other, possibly more significant, benefit. It allows us to record, catalog, and perform statistical analysis on the paths we have experienced.

We will begin with the simplest concept of mapping.

When an organization records malfunction reports (Function – Behavior) with the failure modes (component – condition) that created the situation they need to be careful not to get sloppy in creating the categories. Most systems perform only a half dozen or fewer functions. Also, each function probably only has a half dozen or fewer typical behaviors when they fail. There can be any number of components that might cause failure. Just look through the parts catalog for such components in the installed equipment. On the other hand, it would be unusual to find more than a few components that frequently fail. The terms used to describe failure conditions should also limit the number of component conditions. (It is surprising how many ways can be used to say "broke".) The point being made is that it is absolutely critical to limit the number of characteristics being used to create maps.

Identify the few Functions being done by the system.

Identify the few behaviors observed on failure

Focus on components that have failed in the past and are likely to fail in the future.

Be sparing in terms for component conditions. Do not allow terms that mean the same thing.

For example, if a system performs functions A, B, and C, and function A portrays two behaviors upon failure: 1 and 2, it can be assumed that when an A-1 malfunction occurs, the Failure Modes are reported in the following frequencies:

Component	Condition	Percentage
X	1	80
Y	2	15
Z	1	5

If the data collected follows the pattern described in the table, whenever an A-1 malfunction report is received without additional data, several things can be done:

1. The troubleshooter can be instructed to look for a damaged component X first.
2. If component X can only be replaced by one person or in one shop, it would seem to be a waste of time sending it elsewhere.
3. The inventory can be checked to ensure that component X is in stock.

If all failures and all repairs have been unfailingly tracked and if the system used to record them is designed to avoid using terms that are redundant or realistically impossible, over time, it will be possible to forecast the most likely fix, based only on the malfunction report.

This process actually sounds easier than it is. There are typically a few common practices that throw a monkey wrench into the process:

- People do not discipline their choice of functions, behaviors, and component conditions. As a result, the statistics get all fouled up. For instance if a half dozen different terms are commonly used to describe broke or burnt, the statistics will be distributed among a number of   instances instead of just one.

- Increasingly, craftspeople are really just parts changers who continue changing parts until things start working again. As a result, a perfectly good component is marked "bad order" and discarded. Often, shutting the system down and restarting it, recycling the control com-puter or resetting the connections for control boards, will restart the system. It doesn't solve the problem or identi-fy the defect, but it gets things running again. Once again, the system gets cluttered with mis-information.

It is possible to take the concept of failure mapping a few steps further. Each step requires the same level of discipline described in the basic example above:

1. When a specific Malfunction Report can originate with a component failure in two or more distinct systems, it might be possible to identify "discriminators" that will point to the correct system. Such discriminators can include specific failure codes that are displayed by the control computer, or subtle differences in the behaviors that point to one system and not the other.

2. It may also be possible to identify "quick-fixes" that are associated with certain kinds of malfunctions. In those instances, rather than suffering through an extended outage, it might be possible to simply recycle a com-puter or reset a contactor, get things restarted tem-porarily, and address the defect later when it is more convenient.

As distasteful as it may seem to spend a significant amount of time revisiting the graphic details of failures, there is a significant benefit in knowing what has happened and how frequently each path occurs.

Such an exercise will lead to understanding of:
- The failure mechanisms at work and the clues that need to be watched for.
- The defects associated with the dominant failure modes.
- The paths that have been traveled before and are likely to be traversed again.

With this knowledge and a little hard work and discipline, systems will gain improved reliability and availability.

Chapter 3
Example of Failure Mapping

"We're all capable of mistakes, but I do not care to enlighten you on the mistakes we may or may not have made." **Al Gore**

This chapter is dedicated to providing an example of a realistic problem and how failure mapping can be applied. In this situation as with all examples, the objective of failure mapping is to:

1. Improve the efficiency and effectiveness of repairs.
2. Collect the information needed to improve reliability by eliminating future failures.

As this chapter is perused it is useful for readers to think about how the example fits with their own experiences and how their own situation might be improved if they had a completely functional failure-mapping system in place.

A COLD DAY IN HECK

In the fictional town of Heck, there is a large manufacturing facility that employs a highly- creative staff of engineers and technicians who maintain the systems there. Although these people are good, they are not perfect, and although they do not make a point of openly discussing their failings, they are aware of them and are always looking for ways to improve. The facility also has a central help desk that anyone in the business can call for assistance, and it maintains a DIN (Do-It-Now) team that can be deployed as a first resource to fix any small problems that arise anywhere in the plant or offices.

On one of the first cold days of fall, the central help desk received a call from one of the facilities being served complaining that the office spaces were cold.

Offices across the entire facility were heated using circulating hot water. The hot water in the system was heated with waste heat from a hot oil system that provided heat at a much higher temperature than the hot water, to manufacturing processes used in the facility.

The diagram on the next page provides a general description of the heating system:

The heating system for office spaces and other occupied areas functions as follows:

- Hot water is heated in a shell-and-tube heat exchanger where the water is warmed by absorbing waste heat from a hot oil system that also serves process heating needs throughout the facility. The water is typically heated to around 250-degrees F, but boiling is prevented by maintaining system pressure.
- From the heat exchanger, the hot water is distributed to heating coils throughout the facility. The amount of hot water flowing through any specific coil is regulated by a control valve, and each control valve can be adjusted by a thermostat in the space where the coil is located.
- After passing through the heating coil, water is returned to a water return drum where it is stored before being recirculated.

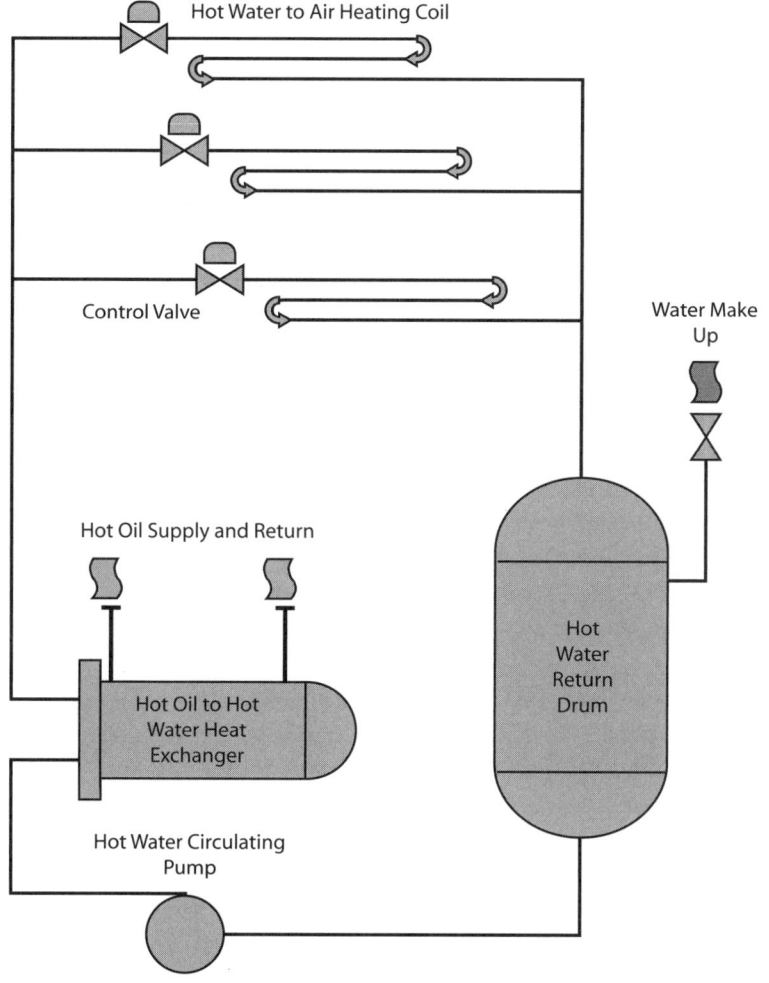

- Based on heating demands, water is drawn from the water drum, and passes through a centrifugal circulating pump where the pressure is increased so that it can once again begin its path through the hot-oil to hot-water heat exchanger.
- The system is operated by the individual assigned to operate one of the manufacturing operations at the facility. The only task this individual normally needs to per-

form is to monitor the water level in the drum and make sure it does not get lower than a prescribed level. The minimum level has been determined by the suction head needed to maintain the Net Positive Suction Head (NPSH) so that the pump does not cavitate. The task is viewed as a nuisance or distraction, compared with other responsibilities that the individual has, associated with the manufacturing operation.

Continuing the fictional account of the events of the cold autumn morning, the help desk operator received the complaint that the office occupied by the maintenance superintendent was

cold when he arrived. As with most organizations, it was not unusual for the maintenance superintendent to be the first person to arrive each morning. This fact was useful to the help desk person because, had someone else made the first report of a cold work space, the help desk operator would have assumed that the maintenance superintendent's office was warm. Otherwise, the superintendent would have complained when he arrived earlier. With this report, the help desk person knew only that one work space was cold.

Checking his failure mapping manual, the help desk person looked up the malfunction report that seemed to most closely relate to the report he had received:

• Workspace HVAC – Not heating

The failure mapping manual is arranged according to Malfunction Reports, then by conditions or characteristics that are useful in funneling, and finally by Failure Modes.

The Malfunction Reports is a list of Functions being provided by the various systems in the facility, with signs and symptoms that are exhibited when the functions are misbehaving.

The characteristics or conditions that are used in funneling can vary widely, based on the sophistication of the systems involved. Computer-controlled systems with self-diagnosing capabilities can provide fault codes that point to a few or even one specific failure mode, based on specific conditions or sensor readings. Less-sophisticated systems also have information that helps in diagnosis of the most likely problem or problems, but this information may be more "manual" or dependent on human sensing and feedback.

The Failure Mode is the combination of a specific component and a component condition that can produce the behaviors being experienced. Obviously, a wide variety of components, and an even wider variety of component conditions, can cause many of the Malfunction Reports that are commonly received. On the other hand, there are only a few "weak links" or components that fail frequently. In addition, for the components that fail frequently, only a few conditions are experienced. In other words, there is a set of Dominant Failure Modes that need to be mapped, and these Dominant Failure Modes provide the information needed to address most of the failures, or at least achieve the two objectives of failure mapping (improve maintenance effectiveness and improve reliability).

Getting back to the fictitious example and the Dominant Failure Modes that will be mapped for this system. (One of the secrets of failure mapping is that it is typically started by connecting Malfunction Reports to Failure Modes. The funneling information that helps connect the two is then added.)

Five Failure Modes are recorded in the failure mapping system:
    1. Thermostat – bad order
    2. Heating Coil – Plugged
    3. Heat Exchanger – Fouled
    4. Pump – Bad Order
    5. Hot Oil System – Out of Service

Based on the reporting history, bad-order thermostats are the most prevalent. First, people are always fooling around with thermostats, so they are frequently mal-adjusted or damaged. Second, the heating coils in working spaces get plugged with dust and debris. This environment in question is used for manufacturing, so the spaces are charged with all kinds of airborne particles. The third most frequent failure mode is a fouled heat exchanger. The make up water in the hot water system is potable or untreated tap water, so it is common for calcium carbonate to be deposited on hot surfaces. Fourth, the only truly dynamic element in the system is the circulating pump, and it has occasional breakdowns. Finally, the hot oil system experiences occasional problems, but it is part of a number of manufacturing processes so it is likely that people would be screaming if it was down for any length of time.

Depending on how long the failure mapping system has existed and how disciplined the users are, it will be possible to go further than simply saying one failure is more frequent than another. It is possible to determine the statistical likelihood of a specific failure mode, based on the number of times it has occurred.

For the sake of discussion, it will be assumed that the fictional facility has had the failure mapping system in place for quite some time, and employees are quite disciplined in its use. Discipline here means asking individuals who report problems to confine their reports to Malfunction Reports or the Specific function that has been impaired, and the specific behavior it is exhibiting. In addition, the individuals who close out repairs are asked to identify the specific Failure Mode. By these rules, the failure mode is restricted to the component that has failed and the specific condition of that component. (If a repair is made by changing out a component but it was impossible to identify the specific aberrant condition, the close out should say "component was changed and function was re-enabled". The correction might have come from recycling the control system or computer, or restoring faulty contacts when the new part was installed. In any event, it is important to help the individual who is performing the investigation of the Failure Mechanism by not providing more information than is known.)

It will be assumed that the following statistics apply:

1. Thermostat – Bad Order – 50%
2. Heating Coil – Plugged – 30%
3. Heat Exchanger – Fouled – 15%
4. Pump – Bad Order – 5%
5. Hot Oil System – Out of Service – 0+%

This is a large facility and it would be time consuming to respond to the wrong failure mode. As a result, it is far better for the help desk person to use some of the information available from where he sits to help funnel down on the real problem rather than to try to gather information by going to the site of the problem or dispatching another person to go there.

The Failure mapping system is specifically designed to assist in this endeavor, and the help desk person is trained to use the failure mapping system as it is designed. The failure mapping system would probably provide the help desk person with the following instructions:

1. Thermostat – bad order – 50%
    a. If a thermostat is BO, the problem is occurring in only one work space. If there are complaints from multiple work spaces, the cause is not likely to be alternative 1 or 2. If several work spaces have filed complaints, go on to alternative 3.
    b. Call the individuals making the complaint and ask themto check if the thermostat is set too low? Ask, if the thermostat setting is raised and lowered can the control valve setting be heard to change, and can the hot water be heard to be flowing through the coil. If no, dispatch a DIN crew person with the appropriate parts. If yes, move on to alternative 2.

2. Heating Coil – Plugged – 30%
    a. Ask the person reporting the problem to carefully place his hand close to the control valve or the supply

line. Is it hot? If so, is there evidence that dust or debris may be plugging the coil? If not hot, dispatch the DIN crew team member with tools needed to clean the coil.

b. If the valve is cold, have the DIN crew member check some of the control valves on coils in adjacent work areas. If they are warm, go back to steps above- and begin troubleshooting for other less-common problems. If they are cold, apply one of the common alternatives listed below that can affect the entire system.

3. Heat Exchanger – Fouled – 15%

a. If there is evidence that the entire hot water heating system is being affected, the most likely cause is a fouled heat exchanger (on the water side). Either check operating logs or ask the DIN crew member to check the discharge temperature of the water leaving the heat exchanger. If the temperature is significantly below design levels, write a work order to clear (hydro-lance) the heat exchanger tubes.

b. If the outlet temperature is at or above normal, check the pump for proper operation.

4. Pump – Bad Order – 5%

a. First, if the person reporting the problem is capable, have him/her check to see if the pump is running. If not, have the reporter try to turn it on. If the breaker trips within a short time, have it reset and try it once more. If it trips again, check that the pump is turning freely. If so, write a work order for an electrician to check the circuit breaker and the circuit for damage.

b. If the pump is turning freely and is operating, have the reporter check the pump by closing the discharge valve and checking the discharge pressure. If the discharge pressure is below normal conditions, write a work order to have the pump inspected and removed and repaired if necessary. If the pump is operating at normal discharge pressure, proceed with troubleshooting for abnormal failure modes.

5. Hot Oil System – Out of Service – 0+%
   Check with manufacturing operations to see if the Hot Oil system is functioning normally. If so, return to steps above. If not, go to the failure mapping system instructions for diagnosing Hot Oil System problems.

The objective of the information provided in the funneling support described above is to assist the help desk person (even one with little experience) to:

- Identify the problem as quickly as possible.
- Use the DIN resources as effectively as possible
- Help triage the use of work order resources (e.g. make most effective use of available resources)
- Make sure repair resources take the proper tools and parts with them to repair sites
- Returning to the fictional account, if the help desk person has access to a mature failure mapping system, he should proceed as below:.

As mentioned earlier, the maintenance superintendent is not normally the first person to arrive on site and the help desk has registered no complaints at that time. So, in a hypothetical situation, how might the help desk person go about determining if he is dealing with a system-wide problem or one that is confined to only one office? By calling up a flow diagram for the entire hot water distribution system, the help desk person can see that the shift foreman's office is served by a different branch of the system than the superintendent's office. So he can call the shift foreman to ask if the heating coil in his office is working. In the hypothetical situation, when asked if the heating coil in his office was warm, the shift foreman responded that it was "as cold as the ————" (fill in your favorite metaphor).

As a result, the help desk person now believes he may be dealing with a system-wide problem.That information helps the help desk person funnel past the first two, most likely causes. While those causes might still be the issue, the situation would require existence of a problem in both the maintenance superintendent's office and the shift foreman's office. If he wanted to further reduce the likelihood of a bad diagnosis, the help desk person could call one or more additional early-arriving individuals to see if their heating systems are working. Let's assume they all respond in the same manner …. "cold as ice".

The next thing the help desk person will do is look for a clue as to the health of the heat exchanger. Several pieces of data are recorded each day on the operating log concerning the conditions in the circulating hot water system, of which one is the water temperature. The help desk person can see that the temperature of the water leaving the heat exchanger is typically 240 to 250 degrees. The most recent report is yesterday, and it is unlikely that the heat exchanger would have failed completely in one day, so the next option is the pump.

While looking at the logs, the help desk person notices that the water level in the Hot Water Return Drum has been chronically operating below the desired level. That fact provides two possible leads to problems, one easy to fix and the other not so easy. The help desk person asks a member of the duty DIN (Do-It-Now) team to go to the heating system location and perform two checks:

1. Fill the Hot Water Return Drum to the normal operating level.

2. If the pump is running, close the discharge valve and check the shut-off head.

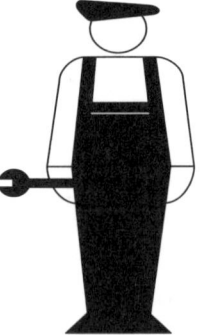

When the mechanic arrives at the work site, he radios back and tells the help desk person that as soon as he walked in he could hear the pump growling and popping. The water in the Hot Water Return Drum was low, and when he filled it to normal level the sounds subsided. When he closed the discharge valve, the discharge pressure of the pump increased only slightly. It appeared that the pump impeller might be damaged.

The sound that the DIN person heard was probably cavitation in the pump. When pumping hot water above boiling temperature, if the appropriate Net Positive Suction Head (NPSH) is not maintained, the pressure at the inlet to the pump will go below the pressure at which water can vaporize.

Vaporization in these conditions will produce steam bubbles that erode the impeller and, over time, will reduce the ability of the pump to achieve the required outlet pressure and so reduce water circulation.

The concept of the Do-It-Now team is to spend the majority of their time making "saves" or doing things that can place systems back in service with short and easy tasks. Only when they have nothing else to do are they assigned to complete longer, more time-consuming tasks. In the situation here discussed, the anticipated repairs were beyond the DIN team's capabilities and the individual who visited the site of the heating system reported his findings back to the help desk person.

At this point, the help desk person had been able to determine that the pump was the likely culprit. Examining the recent maintenance records for the pump showed that it had been overhauled two years earlier and at that time, the pump had a bad seal. The record also had several other comments concerning the condition of the impeller, including references to signs of cavitation, but the impeller had not been changed.

The help desk person wrote a work order to remove and disassemble the pump. Because the actual Failure Mode was still unknown, this step was a part of "troubleshooting" rather than repair. Once the Failure Mode was identified, the work request would need to be updated to include the ultimate corrective steps.

Once the pump was removed and disassembled, the failure mode became apparent. The impeller had experienced severe erosion and was no longer able to pump the desired volume to the required head. The Failure Mode was recorded as "Pump Impeller – Severely Deteriorated".

(The steps described in the last few paragraphs can best be characterized as diagnosis and troubleshooting. In this context, diagnosis is the activity that is completed using externally available information. The more information that is available, and the greater the skills of the diagnostician, the better the diagnosis will be. Also, in this context, troubleshooting is the invasive task of identifying the actual deteriorated component and describing the condition of that component.)

Knowing that the impeller has been severely deteriorated is all the information needed to correct the problem and resume operation …. for this one instance. For an increasing number of companies this is not the stopping point. They want to know what caused the deterioration and why the deterioration was allowed to progress to failure. For those companies, there are two further steps:

1. Failure Analysis
2. Root Cause Analysis

In most instances, the individuals performing the repair are the wrong people to perform these two tasks. One reason is that they do not have time. As soon as they finish one job, they are sent on to the next. Often they also have some "skin in the game". In other words, they may be personally responsible for some of the failures. Finally, they are typically not trained to recognize Failure Mechanisms or to perform Root Cause Analysis.

Let's assume in our example that the facility in question has a reliability engineer assigned to perform Failure Analysis and Root Cause Analysis.

A brief inspection of the impeller and the records of prior pump repairs tell the reliability engineer that the Failure Mechanism was Erosion. The pump had been chronically operated with inadequate NPSH, so the Failure Mechanism was obvious.

The real question was the reason "why". From a review of the pump repair history, it was clear that the impeller had experienced some deterioration before the repair, two years previously. At that time, the decision was made to re-install the deteriorated impeller and to begin operating the return drum at the proper level. For some reason, this solution did not take effect.

Root Cause Analysis is completed at three levels. The first level is the physical cause. The physical cause is most directly linked to the Failure Mechanism. The second level is the human cause. This level is typically when a living, 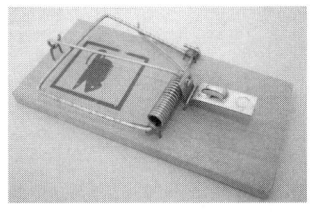 breathing, human being either acts or fails to act and puts the physical cause into action. The final level is the latent or systemic cause, which is the level at which the system or organization creates a trap for the individual to step into. Once a latent or systemic cause is in place, sooner or later an individual will step into the trap and cause a failure. In fact, most latent or systemic causes trap several people and cause numerous failures.

In the present example, the facility had a mechanism for completing failure analysis and for identifying physical causes, but it had no system for identifying human cause or latent cause or for ensuring that solutions were implemented. Believing that causes existed at the physical level only, the facility depended on the capabilities of individuals to see that solutions to physical causes were implemented. If the individuals did not have the appropriate skills needed to implement solutions were not applied. If the primary causes existed at the human or systemic level, there was no mechanism whatsoever for addressing them.

At the level of the physical cause, there were two issues that shared responsibility. First, suction was chronically being maintained at too low a level, causing cavitation and erosion. Second, a worn impeller was reinstalled during the last maintenance event, providing very limited useful life under the current operating situation.

At the level of the human cause, again there are two individuals or roles that shared the responsibility. First, the operating instructions clearly required that the operating level of water in the suction drum be maintained at or above a specific level. The operators who ignored operating instructions and allowed the level to fall too low were at fault. But in addition, the rotating equipment engineer was at fault because he had found that the operating practices were causing deterioration and did nothing about it.

The rotating equipment engineer was a good technical engineer but not so good a project manager. Like many other engineering specialists, he did a good job of identifying solutions. On the other hand, he did a terrible job of implementing those solutions.

At the level of the latent or systemic cause, again there are two shared causes. First, if operations management allows operation of equipment in a manner that causes harm, they are not doing their jobs. After some period, acceptance of poor practices is paramount to endorsement.

Second, organizations need to recognize the strengths and weaknesses of the individuals it employs. If technical experts need a boost in implementing their solutions, the organization should install programs and practices that support those needs. In this example there needed to be a mechanism for the rotating equipment engineer to report evidence of poor operating practices, so that the individuals with administrative authority could make changes.

If the organization were to support Root Cause Analysis down to the human and systemic levels and if it were receptive to the findings of investigations, there would be an opportunity to solve problems such as the example, as well as any others that are likely to fall victim to similar conditions.

Chapter 4
Naturally Occurring Elements ....
That Can Be Enhanced by FM

"Excellent firms don't believe in excellence - only in constant improvement and constant change." **Tom Peters**

In reading through the previous chapters, the reader may have said "There is nothing new in this. We already do this all the time." Although that may be true, it also may not be. Saying it another way, all the elements of Failure Mapping may be in use, but they may be getting done only intermittently, or with so little structure or discipline that only a small portion of the possible benefits are being harvested.

In reading the following sections, a picture of a speedometer should be kept in mind. If it is assumed that the speedometer goes to 100 mph and the job will be to compare the  effectiveness of the current systems to the best possible effectiveness, the speedometer should be allowed to represent that portion of the best possible effectiveness currently being achieved.

Some current systems and disciplines may be harvesting a significant portion of the available performance and the speedometer reading will be close to 100. On the other hand, the current systems may be harvesting only a small portion and the speedometer reading will be closer to zero.

If it is not known how well the current systems are doing, the reader is invited to read the following sections and form a clear picture in his mind of what seems to be a reasonable standard. After forming an opinion of the reasonable standard, the reader should formulate a few questions that can be used to gauge current performance. The reader should ask those questions of the people in the organization who are involved in each element to see how things are going.

IDENTIFYING FAILURE MECHANISMS AT WORK

One of the things that are most astounding to individuals when they begin to focus on reliability in an organized fashion is just how few Failure Mechanisms exist. Up to that point, the forms of deterioration leading to failures seems almost infinite. Few failures occur every day, but their sources seem different from all others and the events leading up to the failures seem separate and distinct from other events.

Once an individual begins to develop an understanding of reliability, he will recognize that there are a few Failure Mechanisms and those mechanisms are parts of a series of events that occur quite consistently over and over again. Although the results may give the appearance of being different, they are much the same. The facts are that: 1) There is a relatively small number of alternatives and 2) There is a consistent pattern that each and every failure follows. These realizations provide the opportunity to convert these strings of events from reactive experience into a pattern that can be managed by using a proactive Work Management Process (WMP).

As with other Work Management Processes, there are two characteristics that separate them from other methods of responding to needs. Those characteristics are:
1. Structure
2. Discipline

Structure – In the context of this discussion, the term structure is intended to imply a consistent organization. A structured process includes a series of steps that are well defined and are applied in a similar manner in each and every situation.

Discipline – In the context of this discussion, discipline is intended to imply behavior that is well defined and controlled within certain limits. In the case of a structured work process, the term discipline would be used to describe how tightly participant behaviors are defined by the process and how closely the participants are required to follow the descriptions.

For purposes of this discussion, the above-defined characteristics will be used to compare "naturally occurring" activities within the typical "failure response" to comparable activities that will exist in a system and work process that is based on Failure Mapping. The objective in making this comparison is to provide the reader with some semblance of the opportunity that is attainable from introducing and using a highly-structured and highly-disciplined Failure Mapping process as part of the method being used to respond to all failures.

Naturally Occurring Activities that are part of Failure Response – In the context of this discussion, this expression is used to describe the activities that individuals perform based on their own knowledge and experience, or steps that have become accepted practice over time because of a long history of accepted behavior. Some of these activities are good, . In others , they are neither effective nor efficient. In either event, they are typically poorly integrated with the needs and capabilities of other systems, processes, or organizations.

STRUCTURES WITHOUT FAILURE MAPPING

Most facilities that have an engineering section of their organization, and those who do not, but have been in existence for quite some time, typically have some activities aimed at identifying Failure Mechanisms that are actively at work to cause deterioration of physical systems. The most common are activities aimed at searching out situations where uniform corrosion is happening as the result of the failure of coating systems. Even those organizations without failure mapping usually have some method of identifying external corrosion.

As was suggested in earlier paragraphs in this chapter, individuals who have not made a study of reliability may not understand all the Failure Mechanisms or how they result in deterioration or failure. A good example of a poorly-understood Failure Mechanism is fatigue. A fatigue failure is the result of a component being exposed to repeated cycles of tensile loads and compressive loads that are beyond the fatigue limit for the material in question. A good example of a component that will rack up a large number of fatigue cycles in a relatively short period of time is a rotating shaft (such as a pump shaft). If the shaft has been poorly aligned, or if the rotating assembly has been inadequately balanced, or if the inlet or outlet nozzles are being distorted due to piping strain, the shaft will experience fatigue. Depending on the load, a failure will occur in some specific number of cycles. If fatigue is not one of the Failure Mechanisms that is recognized and addressed by an organization, fatigue-related failures will occur unabated. It is reasonably likely that an organization without some form of structured failure mapping will not have systems in place to identify more-sophisticated failure mechanisms such as fatigue.

STRUCTURES WITH FAILURE MAPPING

As mentioned earlier, there are relatively small numbers of Failure Mechanisms, but it is important that each and every one of them be well understood by individuals who are responsible for identifying instances where they exist and are causing deterioration.

In a description of the anticipated structures within a process that includes all the elements of Failure Mapping, it would

Failure Mechanisms for Mechanical Devices – The only Failure Mechanisms that can lead to deterioration of mechanical devices are:

Corrosion

Erosion

Fatigue

Overload

Failure Mechanisms for Electric Components and Devices – The Failure Modes for electrical components are:

Overload

Supply Transients

Load Stall

Electrical Analog to Fatigue

Persistent excessive loading at less than levels resulting in immediate breakdown.

Insulation Breakdown

Heat

Chemical exposure

UV exposure

Mechanical Failure

Abrasion

Loosening

be expected that all participants who are in a position to identify Failure Mechanisms at work would be trained and knowledgeable of the Failure Mechanisms described above. These individuals would know where the Failure Mechanisms are likely to occur and they would understand the signs that indicate a specific Failure Mechanism has been at work. A simple example is uniform corrosion. Uniform corrosion of a ferrous metal produces iron oxide or rust. When you see rust, you know that corrosion is at work.

Beyond this simple example, it is possible to teach those same people about the conditions that set the stage for the Failure Mechanism to start working. Using the example of uniform corrosion, there must be an anode, a cathode, and an electrolyte, for a corrosion cell to exist. Two dissimilar metals will provide an anode and a cathode. A leaky door seal that allows water to enter where it should not can provide the electrolyte. If personnel are trained in the conditions to beware of, there is an good chance they will prevent a Failure Mechanism to form before it ever starts working.

The final part of the structure that should exist within a "Failure Mapping environment" is a clearly stated expectation that all individuals participate as active members of the "Failure Mapping Team". In other words, **knowing is not enough**. In addition to understanding Failure Mechanisms and the conditions that create them, these people are expected to take action when their presence is recognized.

DISCIPLINE WITHOUT FAILURE MAPPING

In a "Non-FM" environment, finding an active Failure Mechanism is limited to two possibilities:

1. For those few Failure Mechanisms that are recognized and for which an organized effort has been mounted, it will be possible for the specific individuals who are assigned to the inspection program to identify instances falling into their areas of responsibility.
2. For other forms of Failure Mechanism, and for instances beyond the area covered by the organized program, finding an active Failure Mechanism will be pure happenstance.

DISCIPLINE WITH FAILURE MAPPING

In a "FM" environment, finding an active Failure Mechanism will be based on "control and assurance". As with the old term "QA/QC", in a highly-structured and -disciplined process, it is possible to create protocols that will provide "controls" or activities to ensure that everything is being covered in the desired manner. In addition, it is possible to create oversight activities that provide the assurance that controls are functioning as designed.

When all participants know what to look for and have clear knowledge that looking for Failure Mechanisms is a part of their job, it is possible to ensure that all possible systems and components are being addressed through organization of operator rounds and craft assignments. If any Failure Mechanisms are allowed to proceed unabated to failure, there can be one of two causes:

1. The area where the Failure Mechanism occurred was not on anyone's radar.
2. The individuals who were responsible for the affected areas were not doing their job.

In either event, corrective action is needed.

Reminding the reader of the mental exercise that was recommended at the beginning of the chapter to help determine the value that is possible from implementing a structured and disciplined Failure Mapping process, ask yourself:

- If one of the operators working at any location were to see a component that is rusting away, what action would they take?
- Have all operators been trained to recognize the signs that each kind of Failure Mechanism is at work and the conditions that create Failure Mechanisms?
- Does each operator understand what actions are expected of him when he observes an active Failure Mechanism?
- If a craftsperson found himself reefing on a ten foot cheater to align the piping with a pump housing, would he understand that his actions are likely to create a defecthat will result in fatigue and ultimate failure?
- Are protocols in place that explains to the craftspeople how they should respond when they find such a situation?
- Are assurance systems in place that will verify expectations concerning the identification and elimination of

Failure Mechanisms are being achieved?
• Make an estimate of how many active Failure
  Mechanisms currently exist in the facilities in question.

IDENTIFYING DEFECTS BEFORE THEY CAUSE FAILURES

The practice and process of identifying defects before they cause a failure is much the same as identifying the Failure Mechanisms while at work. The only difference is that, once the defect has formed, the system is one big step closer to failure. An important point to keep in mind is that there is no immediate relationship between the formation of a defect and the occurrence of a failure. Some defects can exist for long periods before the situations occur that result in a failure. In these instances, if the defect can be found before the failure occurs; all the negative consequences of the failure can be avoided. The occurrence of a failure after the defect is formed is akin to throwing the second of a pair of dice after the number on the first has already been determined. Knowing of the existence of defect cuts the odds significantly.

For an example, let's use a situation where a pipe has been exposed to corrosion and has thinned beyond the minimum wall thickness. The minimum wall thickness is the minimum amount of metal needed to contain the fluid at the highest possible pressure. For example, the pressure may be highest when the piping system is blocked-in or dead-headed. If the thinned piping is detected before circumstances result in the maximum pressure level, it will be possible to shut down, drain the system, and replace the thinned piping without a dramatic failure.

STRUCTURE WITHOUT FAILURE MAPPING

In a Non-FM environment, there is little if any structure available to create sensitivity to the presence of defects. Unlike Failure Mechanisms, there is no explicit number of defects. There might be examples of defects that have occurred in specific locations in the past. And if individuals who have dealt with the failures resulting from those defects are still around, they may know what to look for and where.

On the other hand, without the structure associated with FM, there is no formal system for cataloging history and calling it to the minds of the right people in the right places at the right times. As a result, finding a defect in the period between defect formation and failure is a matter of luck.

STRUCTURE WITH FAILURE MAPPING

As failures are mapped, they create a useful history that can be placed in a database and used as a reminder for experienced individuals and as a portal into the past for new employees. With proper attention paid to classification of all important descriptors, it is possible to provide all participants with knowledge of events occurring, not only in their work area but elsewhere, but involving circumstances similar to those that exist in their area.

As with Failure Mechanisms, in addition to having a well-designed database to contain the Failure Maps in a consistent and understandable format, it is important that participants understand that they are responsible for making themselves familiar with the kinds of defects likely to exist in their area of responsibility and that they actively apply themselves in recognizing and identifying those defects.

DISCIPLINE WITHOUT

As with finding a Failure Mechanism without FM, finding a defect is also a matter of luck. The only difference is that the likelihood of finding a defect between the time it forms and the time a failure occurs is much smaller. Think of these differences:

- Failure Mechanisms can work for a long time before a defect results.
- While the Failure Mechanism is at work, the defect does

not exist. Failures result from defects, not Failure
Mechanisms.

- Many Failure Mechanisms are fairly obvious. They are the
  result of specific patterns and they produce results that
  can be observed.
- Many defects remain obscured. To find a defect, you need
  to expect a defect and then look where you expect it to be.

The last difference is probably the most significant with
respect to discipline. Without a well-designed and structured
approach to cataloging them, how is it possible to describe the right
places to look for defects? Individuals with a lot of experience and
an intuitive understanding of how physical systems work may be
able to point to areas where they believe defects may hide. Few of
the individuals who function on that intuitive basis can explain to
others how or why they know what they know. As a result, they can
neither expand their capabilities to others nor make much of a compelling case for action to non-believers.

DISCIPLINE WITH FAILURE MAPPING

If the plant has a system that catalogs failures according to
location, predecessors, related circumstances, etc., it will be possible to identify all similar situations and highlight instances where
individuals should be on the lookout for defects. As an example, in
a refinery, hot-moist gas streams were being mixed with cold-dry
streams. The fact that there were areas below the dew point of the
moisture in the warm stream led to condensation and accompany-
ing erosion. Recognizing that this pattern existed provided engineers with an opportunity to search for all other similar situations in
the refinery. The areas of possible erosion were then monitored for
erosion and resulting defects.

Although this example may be beyond the technical
prowess of many organizations, there are a variety of less complex
instances where Failure Mapping would identify situations and patterns that can be closely monitored. Examples include:

- Formation of cracks after unit cycling in cold weather
- Corrosion under insulation after particularly severe storms
- Defects resulting from atmospheric contamination

In each of the above situations it should be possible to recognize the possibility of more general damage after the first occurrence. A structure response should then be expected by all other people who have had similar experiences and in every instance in similar situations in the future.

In a FM environment it is possible to raise the level of expectations for all events resulting from similar patterns after a failure is first mapped.

To remind the reader of the mental exercise that was recommended at the beginning of this chapter to help determine the value that is possible from implementing a structured and disciplined Failure Mapping process, ask:

If a failure resulted from a specific defect last year, and it was known that the defect had been in existence for quire some time, what are the expectations with respect to that defect in the future?

- Is it to be expected that someone would find the defect before it caused a failure?
- How would the organization react if the defect once again formed, it was not found, and a failure occurred?
- If a failure occurred in one unit and it was known that similar defects were possible in other units, would all other similar instances be identified by the organization?
- How does the organization respond?
- Just guessing, how many defects currently exist in your facilities?

FAILURE RECOGNITION AND DESCRIPTION

In many organizations, Failure Recognition and Description may be the single greatest area of opportunity for several reasons:

1. A clear description of the failure is the starting point for the Failure Map. Think of this condition as if someone was taking a commercial airline trip between two cities in the United States, and think of all the combinations of two cities that could be chosen. Assume that the starting point of the trip is clearly known, the starting point is a large city with several airports, and that the specific airport the person is departing from is also known. Having this information would significantly reduce the number of possibilities. Knowing the specific starting point and all the possible destinations connected to that starting point, it is possible to describe the realistic possibilities. Further, knowing the percentage of total ridership to each destination from the starting point would help narrow the most likely possibilities even further.

2. Creating protocols for clear failure descriptions is relatively easy. Assume that the failure description is to be a Malfunction Report that is a combination of the Function that has failed and a description of the specific behavior being observed. Most systems perform only a handful of functions, and most functions have only a handful of ways they can "misbehave". As a result, it is possible to create drop-downs in databases that provide the user with only the most reasonable alternatives from which to choose. Users cannot input impossible choices or use terms that either the system or other users will not understand.

STRUCTURE WITHOUT FAILURE MAPPING

In a Non-FM environment, individuals are allowed to describe failures in any manner they choose. As a result, it is not known which system Function has been lost or even if a system-function has been lost. Anyone who has had the opportunity to review an assortment of work requests is aware of the variety of forms the failure description can take. Personally, the author can recall instances where the individual filling our the work request simply said "Pump B.O." as a way of telling that a specific pump was "Bad Order". This kind of input provided very little value in understanding what had failed and what the likely response should be. Thinking about this example in the context of a Failure Map, the description provided a very general starting point but no suggestion concerning the direction of travel.

STRUCTURE WITH FAILURE MAPPING

A FM environment has clearly-defined requirements for recognizing and describing failures. This step in the failure mapping process will be accomplished by identifying the affected Function and describing the associated Behavior (or misbehavior). There are several features that make this data particularly helpful:

1. By focusing on a system-function, it is possible to clearly define what level of performance meets requirements and what does not. With this knowledge, individuals are better able to identify situations that can be viewed as "malfunctions". For instance, if an engine is limping along producing 88 percent of maximum horsepower, but still running, it that a malfunction or not? Or can a 12 percent loss even be recognized?

2. As mentioned earlier, a specific system typically has only a handful of functions so each of those functions can be embedded in look-up tables. Also, typically, each function has only a handful of abnormal behaviors, so it is possible to embed those in a look-up table that is associated with a specific function. As a result, the individual describing the failure can only select one of a number of realistic possibilities, and that selection precludes the possibility of using some "off-the-wall" description that provides no help in guiding the response.

3. Each Function-Behavior combination can point to only a handful of possible failure modes. Many of those alternatives can contain direction-finding assistance at the "funneling" point, generated by computer fault codes or other information that is part of the failure pattern.

To summarize, there is far more structure in a FM environment and the way that structure is designed will point the way toward the actual problem and the most likely solution.

DISCIPLINE WITHOUT FAILURE MAPPING

The discipline possible in a non-FM environment is based on continual effort. Someone will need to monitor work orders every day to see that individuals are doing a good and thorough job of identifying failures and describing the failure, and take action to remedy the situation

Take the example described above. There is a worst possible situation than writing a work request saying, "Pump B.O.". The worst possible situation is when no one writes a work request at all. The disabled spare pump is then not discovered until the companion pump also fails, when the entire system or unit or plant is shut down. If situations like this are unfamiliar to the reader it is wise to begin looking for a failure mode resulting from a human cause that could have been avoided and for which someone feels responsible. Despite that insight, there is limited knowledge of the starting point and even less information concerning the direction of response.

The message is that without FM there is typically little structure and without structure, there can be no discipline.

DISCIPLINE WITH FAILURE MAPPING

As soon as a reasonable amount of failure mapping has been completed, there will be a growing knowledge of the relatively small number of Malfunction Reports (Function – Behavior) that need to be built into the system. Once the failure reporting system has had the appropriate Functions and Behaviors embedded into look-up tables, the system becomes "self-disciplining". People simply have to select from the available choices those that most closely describe the current situation.

The reader may think that the difficulty of this step is being understated, but that is not so. If the reader is familiar with RCM (Reliability Centered Maintenance) or any other form of FMEA (Failure Modes and Effects Analysis), he will realize that typical systems have only a handful of functions that need to be protected.

There is greater difficulty in identifying the small number of terms to be used to describe behaviors because individuals are typically so creative in selecting descriptions, and all the subtle nuances that a behavior might exhibit. Take for example the possible choices when a device has overheated. Typical choices that people may select, if they are not restrained to a single choice are:

- Overheated
- Burned
- Burned Up
- Cooked
- Fried
- Melted
- Charred

Each term represents much the same condition and the same starting point for a failure map. If individuals who are reporting the failure are not directed to use one, and only one of these choices (overheated for instance), each instance will create the starting point for one or more maps and disrupt the usefulness of the maps and the statistical analysis that results.

The point is that, in designing the basis for a failure mapping system, it is best to be early and thorough in identifying behaviors. Behavior descriptions that are nearly the same should be eliminated.

To remind the reader of the mental exercise that was recommended at the beginning of the chapter, to help determine the
value that is possible from implementing a structured and disciplined Failure Mapping process, review a reasonable number of
current work orders and then answer the following questions:

- Does the Malfunction Report identify the affected function and the aberrant behavior?
- Does the work order describe a condition that it would
  be impossible for the writer to understand? (e.g. Would
  X- ray vision be needed to see an internal defect?)
- Does the work order request work to be done that the
  writer may not be absolutely certain is needed?
- Is it possible for the expected response to this work
  order to be misinterpreted resulting in unnecessary work
  being done?
- How is feedback provided to the writer when a work
  order has been completed improperly?

DIAGNOSTICS

**An important point that concerns this and the next few
sections (Diagnostics, Funneling and Troubleshooting) is that
all of these events always happen. Although that probably
doesn't seem like a big deal, it can be. These steps can occur
in the right setting or the wrong setting. They can be done by
the right individual or the wrong individual. They can be done
with the right information or without it. If someone does not
perform diagnostics and funneling using failure maps and all
the information that is available externally, the individual performing the troubleshooting will do so by tearing things apart.
This alternative can turn out to be an expensive and time consuming approach. It always happens and it is up to the organization how it will happen.**

In the context of this discussion, diagnostics is intended to
refer to the analysis of the failure based on externally-available,

even remotely-available, information. At the conclusion of the diagnosis step in a mature reliability process, the diagnostic technician will be able to do several things:

1. The defect that caused the failure will be known.

2. It will be possible to perform triage, identifying the best way to address the problem from among several alternatives and how the urgency and ability to solve the problem fits compared with several other problems currently being addressed.

3. The best first approach for the troubleshooter to take so that time and resources spent on false starts will be minimized will be established.

Some organizations have specific individuals in specific roles, who are trained, equipped, and provided access to information needed, to make them effective as diagnosticians. Other firms do not. As mentioned above, all organizations must perform the task of providing diagnostics. Although some do it effectively, others have craftspeople perform diagnostics, with the whole world waiting on them to figure out what is wrong, and with very little help.

STRUCTURE WITHOUT FAILURE MAPPING

The structure of diagnostics without failure mapping can range from fairly sophisticated to non-existent. At the sophisticated end of the spectrum, knowledgeable individuals spend time studying the available evidence, then make recommendations on how repair efforts should proceed. This approach is good, but it is time consuming and highly dependent on the capability of the diagnostician. On the "non-existent" end of the spectrum, the repair task is assigned to a craftsperson, who proceeds to identify the defect by disassembling equipment. This approach also can be both time consuming and expensive.

STRUCTURE WITH FAILURE MAPPING

In a FM environment, a database exists that contains a set of "maps" showing all the paths that can result from each Malfunction Report. For the sake of discussion it will be assumed that one Malfunction Report has six possible Failure Modes, each of which could produce the same Behavior. Also for the sake of discussion, it's assumed that the six possible modes can exist in two different systems and two different subsystems in each subsystem. It's also assumed that the practice of failure mapping has been in place long enough for a large enough number of instances to have occurred to populate the statistics needed to identify which of the failure modes are most likely and which are less likely to have occurred.

In graphical form, the hierarchy of choices would look as follows:

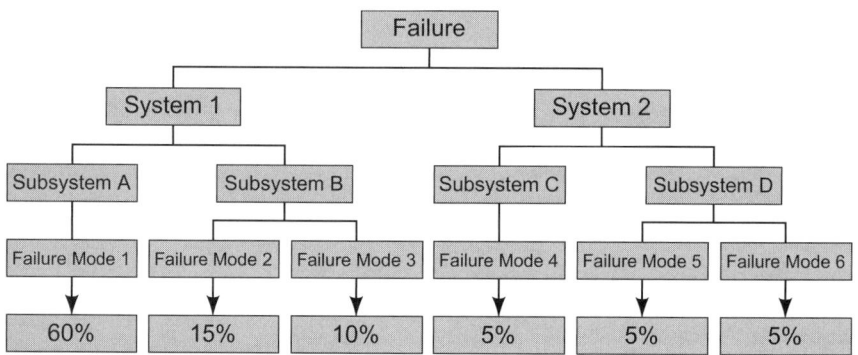

In this example, the diagnostician could quickly arrive at the fact that the greatest likelihood is that the defect is Failure Mode 1, based on the fact that 60 percent of all past failures reported using the specific Malfunction Report have been mapped to that cause.

Depending on the role of the diagnostician and the structure of the organization, the diagnostician may also recommend that the initial approach to repair be accomplished by a specific resource that is best equipped to attack Failure Mode 1.

In another situation, where a graphical representation of the failure map looks as follows:

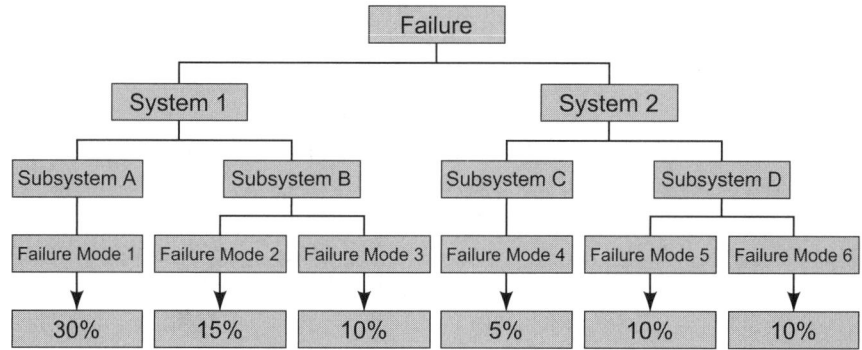

The diagnostician may need to ask some questions or collect some data that would aid in "funneling". The two distinct systems shown here are equally likely to contain the defect. To achieve a more accurate and useful diagnosis, it will be necessary to gain some additional evidence that points to either one system or the other. A more-thorough description of alternatives for funneling is provided in the next section.

DISCIPLINE WITHOUT FAILURE MAPPING

In a non-FM environment, the discipline that ensures that someone performs diagnosis before tearing into a piece of equipment looking for the defect depends on the character of the person performing the diagnosis. If there is no person performing diagnosis, there is no discipline. On the other hand, if the individual has a reputation of being able to quickly identify a defect and by doing so, save a great deal of time and resources, the organization will pay homage to that talent and demand participation before taking any further steps. Of course, that eventuality depends on the skilled person being available to perform the diagnosis. The capability then depends on a person and not a system.

DISCIPLINE WITH FAILURE MAPPING

In an FM environment, all the information in the system is easy to use and available to any number of people. For that matter, any number of people can use the information at the same time, so if there are multiple failures no-one needs to wait on anyone else.

As a result, the organization can reasonably expect that each and every failure will be diagnosed prior to sending out any workforce to begin troubleshooting or disassembly. This characteristic creates a significant benefit for organizations that lack depth, or one that is in transition from having lots of people with strong work histories and experience to a younger, less-experienced workforce.

To remind the reader of the mental exercise that was recommended at the beginning of the chapter, to help determine the value that can be derived from implementing a structured and disciplined Failure Mapping process, ask:

- In this your organization, who performs the diagnoses?
- Are the diagnoses performed before individuals are assigned to undertake the repair or after?
- Are resources being wasted by the way the system currently performs diagnoses?
- For what proportion of the time is the current diagnosis correct?
- How often does troubleshooting end up being a wild-goose-hunt?
- What proportion of failures has a clear understanding of the Failure Mode that caused the failure?
- If one person or a few individuals are expected to act as diagnostic experts, what happens when they are off work? How close are they to retirement? What is the condition of their hearts and other physical attributes?

FUNNELING

In the poorly-defined region between diagnostics and troubleshooting is a step called funneling. Sometimes this step is needed, sometimes it is not. In the example above, there is a fairly straight path between the Malfunction Report and the most likely Failure Mode. In other instances, the path is not so direct.

When the path is unsure, there is a possibility that one likely Failure Mode is possible in one system or subsystem, and another likely Failure Mode exists in another system or subsystem. In this situation, there is a need to look for clues pointing to one Failure Mode or the other. In many situations, there are subtle differences in the behavior. In others, there are "fault codes" that are available from computer controls. In still other instances, it is necessary for someone to look for physical evidence like dripping oil, or signs of overheating.

In any event, someone needs to know what to look for specifically and to ask the person best able to answer the questions. That person may be the equipment operator who is closest to the installation and is best able to answer. The diagnostic technician may ask the questions needed to accomplish the funneling step or it might be a work planner. However, the questions must be asked and answered before triage can be completed or craftspeople are dispatched to begin disassembling equipment.

STRUCTURE WITHOUT FAILURE MAPPING

Attempting to describe how funneling is structured in an environment without failure mapping is difficult. It is like attempting to describe the structure that would be used for describing directions for traveling from one place to another in an environment that had no maps and no directions. There would be no consistent structure. Everyone would do it differently.

When a FM process has been installed and the organized effort of describing the paths from Malfunction Report to Failure Mode has started, the analogy will become more apparent. In many situations, an analyst will look at the initial Malfunction Report and the final Failure Mode that was used to close out a repair job and wonder how a mechanic ever worked his way through the maze to solve the problem. In the real-life situation, where there is no apparent link between the Malfunction Report and the Failure Mode, it must have taken a great deal of time and trial-and-error-work to find defects.

The best way to describe funneling without FM is just that trial-and-error-work. With no landmarks or intermediate directions between beginning and end, it is up to the skills and knowledge of the person performing the repair to identify the clues, solve the puzzle, and find the defect.

STRUCTURE WITH FAILURE MAPPING

As with guide posts and road maps showing directions, funneling queries and comments tend to provide intermediate clues pointing toward one end-point or another in a mature FM environment. Individuals sometimes intuitively understand the clues being provided by FM funneling without access to a structured failure map, but those are exceptions. FM provides the structure needed to capture the clues and make them a permanent part of a system, so that success is less dependent on the skills of individuals and more dependent on the capabilities of the organization.

DISCIPLINE WITHOUT FAILURE MAPPING

In a non-FM environment, there are two possible alternatives when a person is given a Malfunction Report to address:

1. There is no information about the Failure Mode, so a broad search must be started to find the defect.
2. There is some information about the several Failure Modes that may be present, but nothing that points to which is the most likely.

As a result, a limited amount of discipline can be expected. People act based on either experience or intuition. In other words, if they have made a similar repair in the past, they begin by trying what worked before. If applicable experience is lacking, the person acts on intuition, which may be called something else, but intuition is the appropriate characteristic. In either situation, there is no reasonable basis for any expectation that the job will be completed in a timely manner.

DISCIPLINE WITH FAILURE MAPPING

In a mature FM environment, the person performing the diagnostics has access to a file that tells which Failure Modes have been associated with specific Malfunction Reports. In addition, if there are additional clues or symptoms that point to one Failure Mode in preference to another, the information needed to use those clues is included.

In the mature FM environment, an individual files a Malfunction Report. Upon receiving the Malfunction Report, the diagnostician looks up the specific Malfunction Report in the FM file. If there are several possible Failure Modes, the diagnostician looks at what information is needed to "funnel" down to the most likely Failure Mode. In some situations, the data might be an alert code issued by the computer control. In others, it might be some additional symptoms. In still others, the additional clues might come from recent maintenance history or other data. If the funneling information is available remotely (without touching the equipment), it is expected that the diagnostician will gather the data and issue troubleshooting instructions pointing at the most likely candidate.

The discipline in a mature FM environment comes in the expectation for quick and accurate solutions to problems that historically have seemed vague and difficult to identify.

The reader should be reminded of the mental exercise that was recommended at the beginning of the chapter to help determine the value that is possible from implementing a structured and disciplined Failure Mapping process, so ask:

- Does the current computer control system provide fault codes that are unused?
- How is the true cause isolated when several sources can have caused a failure?

TRIAGE

Many people will recall watching the old Korean War show MASH on television or other shows featuring the activities around hospital emergency rooms, and will have heard the term triage. In these situations, triage is the process of pre-screening patients for several characteristics to determine the urgency for use of limited resources. Triage asks which of the patients:

- have injuries or illnesses that can wait for care?
- need immediate assistance?
- need some form of specialized care now?
- are unlikely to survive, independent of the level of care or the time when it is provided?

The same kind of sorting activity is valuable when dealing with problems involving physical equipment. Most organizations depend on their prioritization process to determine where the help goes first and how available resources are distributed. On the other hand, if you know how every Malfunction Report is likely to be resolved, it is possible to apply the right resources in the right sequence.

STRUCTURE WITHOUT FAILURE MAPPING

If triage is accomplished in a non-FM environment, it is based on the severity of the Malfunction and the politics of the specific organization. For instance, if a relatively minor problem results in the loss of a major asset, a significant portion of the available resources will be dispatched to deal with that problem. In other words, the response will be based on the appearance of severity rather than the actual severity. It is possible to build a more complex and sophisticated system for prioritizing the response and responding, but all enhancements are like "putting lipstick on a pig". Without knowing how the issue is likely to be resolved, any enhancement is still based on guess-work.

STRUCTURE WITH FAILURE MAPPING

In an FM environment, planners, schedulers, and troubleshooters can prepare to use their time and resources to deal with the most likely Failure Modes associated with all the accumulated Malfunction Reports. Even high-visibility failures that require minimal effort to repair will attract only the attention and resources they need.

Some maintenance organizations have a variety of different approaches to perform maintenance. Some of these include:

- DIN (Do-It-Now) crews that are equipped to handle light maintenance on-site.
- Help Desks that are organized to provide advice that operators can use to correct problems themselves.
- Servicing Facilities that are capable of light repairs.
- Shops that are capable of most normal repairs.
- "Back shops" that are capable of major repairs and overhauls.

In an FM environment there is an expectation that the diagnostician performs triage and assigns the work to the proper source for correcting the most-likely Failure Mode.

DISCIPLINE WITHOUT FAILURE MAPPING

In a non-FM environment, there is little triage. The only discipline is that the work is assigned first to the least-costly resource and sequentially to increasingly more-expensive or scarcer resources. This discipline is one of economy rather than of effectiveness. The approach saves costly and scarce resources but does so by sacrificing time. The only other alternative is the "squeaky wheel gets the grease" approach. Problems with the loudest or most powerful sponsor get use of the best resources first.

DISCIPLINE WITH FAILURE MAPPING

In an FM environment, it should be expected that two things seldom occur:

1. A task should seldom need to be sent to another resource because the first one was unable to address the problem.

2. A resource should seldom complete a repair, and then say "That repair could have been made by a less-costly method or less-scarce resource."

FM provides the knowledge needed to make the right choices to fix things in the least time by the least costly resource.

The reader should be reminded of the mental exercise that was recommended at the beginning of the chapter to help determine the value that is possible from implementing a structured and disciplined Failure Mapping process. Ask:

- What is the least expensive way to complete a repair in this organization?
- How many different alternatives for completing a repair exist in the organization?
- How are problems now matched with the quickest and least expensive method of repair?
- When there are a number of instances needing emergency response, how is it decided which to attack first? Is the decision based purely on importance, or on which can be repaired most quickly? If the latter, how is that determination made?
- Does the organization have clear protocols and procedures for performing triage?
- Who performs the triage and what tools are used?

TROUBLESHOOTING

In the context of this discussion, troubleshooting is a "hands-on", frequently invasive activity, that is time consuming and expensive. It is the step needed to confirm the identity of the defective component. Historically the term troubleshooting has been used in association with an activity that involves tearing into all the components that may be connected with the failure that was experienced.

Occasionally, individuals and organizations get good or get "lucky" at troubleshooting because they have experience with similar failures and simply replicate their successful efforts. In this situation, the organizations have a kind of informal Failure Mapping because the failure maps become imprinted on the memories of the individuals who are involved. The weakness with this approach is that people's memories are a particularly weak tool. People tend to remember good experiences and forget bad ones. They also tend to be reassigned or retire, taking all their knowledge with them. There is no system to this approach that can be recalled in an organized fashion. The memories do not appear on cue when needed.

STRUCTURE WITHOUT FAILURE MAPPING

Troubleshooting without FM is a hit-or-miss activity. An OEM manual may provide a list of things to check when certain symptoms are experienced. That list will help the troubleshooter find the most obvious and common kinds of failures. On the other hand, the OEM manuals are typically written before a product is released or early in its life, before most of the Failure Modes have started to announce their existence. As a result, OEM manuals are of limited value.

Another weakness of OEM manuals is the context in which they are written. First, they are not really written with consideration of the exact application of the equipment. Many Failure Modes are associated with a specific pattern of use. Second, OEM manuals have no way to know the organization and skills of the available resources, so they cannot speak to the kinds and levels of troubleshooting that can be expected in differing places.

There can be little meaningful structure without adapting troubleshooting guidelines to specific organizations and Failure Modes, as will happen when implementing FM.

STRUCTURE WITH FAILURE MAPPING

In a mature FM environment there is a list of all possible Failure Modes that can be associated with any Malfunction Report. Over time, history creates a record of the relative likelihood of each type of failure, based on the number of times it has occurred. If there is a question concerning which of the Failure Modes is most likely, there are frequently other clues in symptoms or control computer messages that point to one Failure Mode or another.

This information provides the person performing the troubleshooting with clear directions on where to look for failure-causing defects.

DISCIPLINE WITHOUT FAILURE MAPPING

Without FM, troubleshooting expertise is based on skill and intuition. When thinking about how discipline might be provide to the organization to enhance effectiveness, there is little "footing". Typically individuals who specialize in troubleshooting are chosen because they can identify find problems more quickly than simply grinding from one alternative to the next. These individuals have a special talent that is difficult to quantify, and is almost impossible to build into a course of training.

As a result, how is it possible to determine when someone is doing troubleshooting well and someone is not? How can individuals who are trying hard and those who are not be separated? How are individuals whose intuition is working well to be distinguished from those whose intuition is not working well?

These questions may sound a little odd but they tend to reflect the difficulty of providing any discipline to an unstructured troubleshooting environment. Putting it the simplest way possible, without added structure, "you take what you get". If your troubleshooters find problems quickly, you should consider yourself fortunate, because it is not something that you deserve or have a right to expect. And this good fortune is likely to collapse at the most inopportune time.

DISCIPLINE WITH FAILURE MAPPING

In addition to the structure described above, troubleshooting in a FM environment provides ordered direction as to which Failure Mode should be explored first and the sequential order that should be followed until the defect is found.

In addition to finding the defect more quickly because the attack sequence is according to likelihood, the concept of troubleshooting in a structured order has another benefit. It stops people from spending a lot of time talking and asking valueless questions. Individuals can simply attack the problem and keep working in an ordered sequence until they find the problem. Even if the defect is something other than one of the known Failure Modes, the troubleshooter is likely to find it in the minimum time by just exploring the most likely areas.

To remind the reader of the mental exercise that was recommended at the beginning of this chapter to help determine the value that is possible from implementing a structured and disciplined Failure Mapping process, ask:

- Who performs troubleshooting in the organization?
- What information do the troubleshooters have when they start?
- If a number of different troubleshooting experiences is monitored, what proportion of the time does the troubleshooter attack the right thing first?

- Are diagnostics and troubleshooting well-established concepts? Who is assigned to perform each? What is their relationship to each other?

PERMANENT REPAIR

The concept of a permanent repair is one that is often mistakenly taken for granted. People assume that the repair being applied will address the problem and eliminate the defect. In fact, that viewpoint is quite naïve. The frequency of non-permanent (unsuccessful) repairs ranges from 25 percent to one-third. In these situations, the person who performed the repair took a step that temporarily restored operation, but did not correct the real problem.

Work accomplished in a FM environment has the distinct advantage of requiring a Failure Mode report to close the job. In other words, the person performing the repair needs to provide a failed component and a description of a condition that is consistent with the reported behavior. In the situations described above, where non-permanent repairs are completed, it is common to find that, when analyzed, the removed component had no defects.

STRUCTURE WITHOUT FAILURE MAPPING

Clearly, the concept of a permanent repair can be built into any repair process, even those without FM. It is only necessary to demand that a defective component be produced at the conclusion of the repair. Many intelligent car owners do this when they take their vehicle into a shop for repairs. That way they at least have some idea that the repair was completed, or they may have physical evidence it was not (e.g. the defective part produced is from a different make or model).

On the other hand, it will make more sense to the individuals performing the repair and will avoid the issue of mistrust, if collecting and analyzing defective parts is merely a step in a comprehensive and uniformly-applied process. If participants know that the component is being saved so that Failure Analysis can be performed and the Failure Mechanism identified, the process will make more sense.

STRUCTURE WITH FAILURE MAPPING

In a mature FM environment, there is an expectation that the troubleshooter will keep looking until a component with a defect is found. This expectation is not a random occurrence, it is applied consistently. With this expectation in place, if troubleshooters do not have the tools to identify defects, they will demand that they be provided. Heightened expectations associated with FM will improve results and performance.

DISCIPLINE WITHOUT FAILURE MAPPING

As with the structure as it is described in written procedures and implemented through training, the discipline or actual practice of expecting a permanent repair can be applied in a non-FM environment as well as it can in an FM environment. The issue is not "can you make it work". The issue is one associated with "push" versus "pull". In other words which way is more difficult to make the process work.

Without FM, you will obtain defective parts by demanding them and regularly providing oversight and discipline to the systems that are designed to collect and manage them. In other words, it requires "push" to make these systems function.

DISCIPLINE WITH FAILURE MAPPING

In a mature FM environment, the system itself provides the "pull", or the positive influence to see that defective components are collected and analyzed. By accumulating this data, participants in the process know they will make their own lives easier in the future by improving the completeness of the FM file and the effectiveness of the maintenance process by providing the information that will eliminate defects in the future.

To remind the reader of the mental exercise that was recommended at the beginning of the chapter to help determine the
value that is possible from implementing a structured and disciplined Failure Mapping process, ask:

- What is the established expectation concerning repairs
  in the organization? Are people expected to perform
  permanent repairs 100 percent of the time, or is it
  acceptable to just patch things up and return to service?
- What portion of the time is the actual defect that caused
  the failure not found? Make a list of examples?
- How would the organization react if the same failure
  occurred again?... and again?

FAILURE ANALYSIS

In the context of this discussion and in the structure of the
FM approach described herein, Failure Analysis is the step involving identification of the Failure Mechanism that resulted in the deterioration leading to the defect causing the failure. Although the
Failure Mechanism is one of the steps in a cause-effect-cause-
effect-cause-effect- . . . sequence, it should not be confused with
one of the root causes. Failure Mechanisms are simply nature acting as nature will. For instance, if a situation is created in which two
dissimilar metals are connected by a dielectric, a corrosion cell will
exist and corrosion will occur. That is how nature works and it is
one of the most common Failure Mechanisms. On the other hand,
the cause is whatever and whoever allowed the corrosion cell to be
formed.

Without knowing the Failure Mechanism, it will be impossible to identify the cause and apply some form of prevention.

STRUCTURE WITHOUT FAILURE MAPPING

In a non-FM environment, identification of the Failure
Mechanism causing deterioration that leads to any failure-causing

defect is a hit-or-miss proposition. If the problem created by the failure is sufficiently troublesome, the time and resources needed to identify the Failure Mechanism will be spent. In most situations, those expenditures do not happen, so the source of deterioration goes undiscovered.

STRUCTURE WITH FAILURE MAPPING

In a mature FM environment, identification of the Failure Mechanism is an ordinary part of the process. As a result, two very important things begin to happen:

Failure Mechanisms are continuously being identified and discussed, so the concept is not confined to engineers and reliability experts. It begins to penetrate the entire organization.

More people understand Failure Mechanisms and their causes, so they begin to recognize situations that allow Failure Mechanisms to exist. An environment is thus provided where a much larger group of people can identify and eliminate Failure Mechanisms.

DISCIPLINE WITHOUT FAILURE MAPPING

In the absence of the processes and procedures associated with FM, there is no requirement that Failure Mechanisms be routinely identified. As a result, the existence of active Failure Mechanisms throughout a facility should be expected. There is no basis for any other expectation.

DISCIPLINE WITH FAILURE MAPPING

In a mature FM environment, there is a foundation upon which very high-level expectations can be based. For instance, when a much larger proportion of the population understands Failure Mechanisms, there can be an expectation that they are actively engaged in finding and reporting them. All employees become engaged as knowledgeable participants in the corrective action and continuous improvement process.

To remind the reader of the mental exercise that was recommended at the beginning of the chapter to help determine the
value that is possible from implementing a structured and disciplined Failure Mapping process, ask:

- Is anyone assigned to perform failure analysis and
  identify Failure Mechanisms?
- For what portion of failures is failure analysis completed
  and the Failure Mechanism identified?
- Review at least ten recent significant failures. What
  portion of those have the Failure Mechanism identified?
  Is this number consistent with what is expected in the
  organization?
- Does the organization perform Pareto analysis of the
  Failure Mechanisms being identified? If so, what is the
  most common Failure Mechanism?
- What steps are being taken to eliminate the most
  common Failure Mechanism?

ROOT CAUSE ANALYSIS

In the context of this discussion, Root Cause Analysis is the
proactive effort intended to determine the cause that allowed the
Failure Mechanism to become active at three distinct levels, which
are:
1. The Physical Cause – or physical gap that allowed the
   Failure Mechanism to begin.

2. The Human Cause – or the specific individual whose
   actions or inactions created the physical cause.

3. The Latent or Systemic Cause – or trap in organization,
   procedures, practices, etc. that allowed the human to act
   in the manner that resulted in the physical cause.

STRUCTURE WITHOUT FAILURE MAPPING

As with Failure Analysis, Root Cause Analysis is a hit-or-miss process in most non-FM environments. Frequently root cause analysis is so cumbersome that it is used on a limited basis. Applying root cause analysis infrequently, allows it to remain cumbersome.

STRUCTURE WITH FAILURE MAPPING

When Root Cause Analysis is applied regularly and uniformly, two things will happen:

Root Cause Analysis will become less cumbersome to conduct. The systems must be designed to provide answers quickly and efficiently.

People will become less terrified of being identified as a "Human Cause" and will accept opportunities to change their behavior and eliminate problems resulting from their own action or inaction.

DISCIPLINE WITHOUT FAILURE MAPPING

In a non-FM environment, people remain leery of investigations and resist drilling down to find Human Causes and Systemic Causes. If Root Cause Analysis is reserved for instances where major losses have resulted, no one will want to participate for fear of becoming the "goat".

DISCIPLINE WITH FAILURE MAPPING

In a mature FM environment, sooner or later everyone is identified as a Human Cause and every manager has some part of his responsibilities identified as a Systemic Cause. These events all happen, and people learn from them, changing the way they do things. And it all happens without the world coming to an end.

Over time, more and more failures are mapped, which makes the maintenance process increasingly effective. More causes and Failure Mechanisms are identified and eliminated, raising the level of reliability. When conditions get better, people provide more cooperative and enthusiastic support. Discipline takes care of itself.

To remind the reader of the mental exercise that was recommended at the beginning of the chapter, to help determine the value that is possible from implementing a structured and disciplined Failure Mapping process, ask:

- Who in the organization is assigned the responsibility of performing Root Cause Analysis?
- What proportion of failures is exposed to RCA?
- Is a Pareto analysis of the common root causes performed?
- What are the most common systemic causes?
- What is being done to address those systemic causes?
- Specifically, who were the people named as being the human cause in the last few failures?
- How were those causes addressed?

Chapter 5
Structured Assessment

Those who do not read criticism will rarely merit to be criticized.
Isaac Disraeli

The quotation above is quite apt as an introduction for this section. Although it was intended to apply to criticisms of written works, the quotation is applicable to a much broader array of subjects. Organizations and individuals who do not care what others say of their work will frequently act in a manner that reflects that disdain. If a person asks for an assessment and pays attention to the results, it says that person cares. On the other hand, if an assessment is requested or one is allowed to occur at the request of another, and there is then no response to the findings, that says the person wants to give the appearance of caring. The author prefers to deal with individuals who care, but prefers those who do not care to those who would only like to make believe that they care.

Before discussing details of the specific assessment to be used to determine the value that will result from implementation of the Failure Mapping process in a business, it is proposed to discuss the elements of a structured assessment in general. This level of detail is intended to help the reader understand why a detailed assessment is important. Many readers who are familiar with reliability programs will read this book and simply say, "This approach will help us" or "We are already reaping many of the benefits available from Failure Mapping". On the other hand, many others are fairly naïve concerning their current knowledge of failure patterns and what is possible by addition of structure and discipline.

For these latter individuals it will be helpful to provide a fairly comprehensive look at how the activities that will be enhanced by Failure Mapping are currently being accomplished. It will also be a benefit if those enhancements are translated into the dollars and cents they are currently costing and the amount of improvement it would be reasonable to expect.

The elements of any structured assessment will first be described in this chapter, and the next chapter will further describe the elements of a structured assessment aimed specifically at evaluating the value of Failure Mapping.

Generally speaking, a structured assessment is intended to answer two questions:

1. What are the envisioned changes worth?
2. What will the envisioned changes cost?

To a large extent, it doesn't matter what issue is being considered because the same questions apply. Right off the bat, people want to know what any activity is worth. Maybe a better way of describing this question is the following, "Is this project even worth my time and attention? If not, don't bother me."

If the company can be convinced that the program being discussed is sufficiently valuable to justify their time and attention, then the question expands to one of comparing current performance to the performance that can be achieved after the intended change is complete and calculating the value of the improvement.

After having calculated "what it is worth" and assuming that the audience has remained interested, then the next question to answer is, "What does it cost?" It seems that people hate going over budget even more than not receiving what they were promised. As a result, it is critical that the assessment include all elements needed to make the proposed change completely functional. This total includes both direct costs and the other resources that will be needed. The list of resources may include participation of staff members, office space, software development or modifications, etc.

With the above thoughts in mind, each of the various elements will now be discussed in greater detail, starting with the dia-

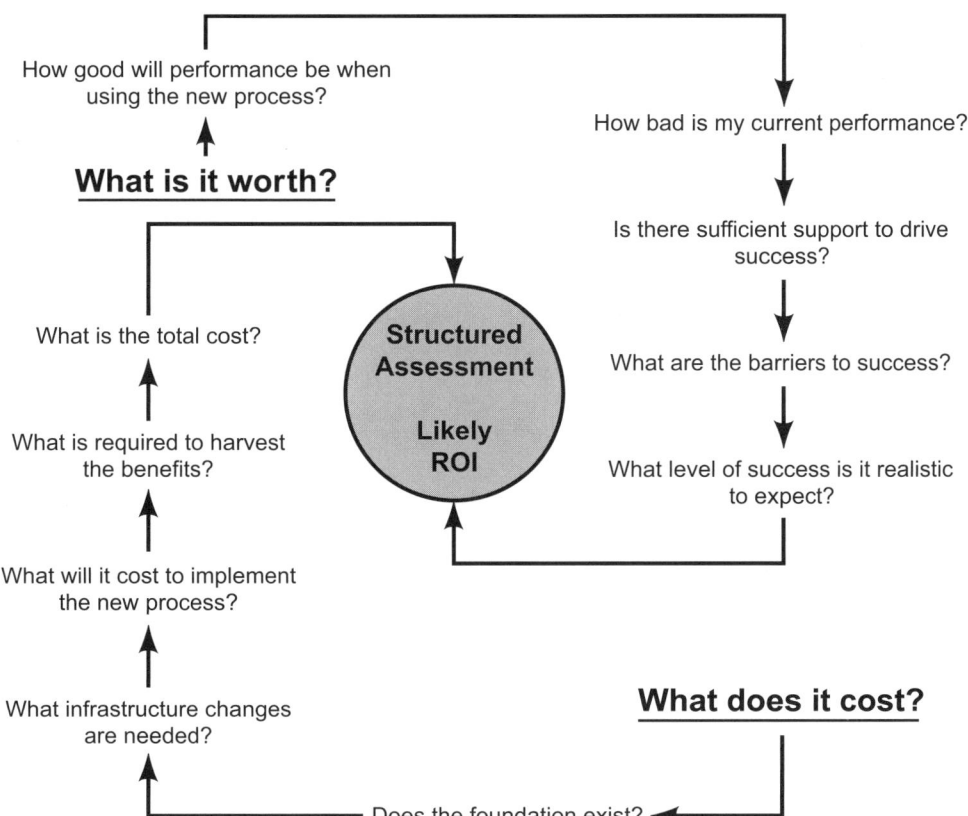

gram above, which is intended to describe the general process used to perform an assessment.

WHAT IS IT WORTH?

The assessment process begins with the question, "What is it worth?" This question expresses the single most-important aspect of the assessment, because it is the one and only opportunity to sell the ideas. If this opportunity is missed, the game is over. The presenter will get to sit down on the bench and watch those who follow to see if they can do a better job.

In selling any change, the presenter needs to be able to paint a complete and compelling picture of the value of the changes that will result from the proposals. It can never be taken for granted that other people can envision all the benefits seen by the presenter. Quite often, the audience sees only what is on the surface and not the "downstream" benefits.

Take for instance, the example of RCM analysis. RCM (Reliability Centered Maintenance) is an analytical technique that identifies the optimum program of predictive and preventive maintenance. From RCM analysis, most people expect the following results:

- Improved predictive and preventive maintenance that will improve reliability by intervening before failures can take place.

While that statement is true, it significantly "undersells" the value that can be achieved from a program that begins with performing RCM analysis. In addition to improved reliability, RCM can be used to either produce or help facilitate the following:

1. The program of Predictive and Preventive Maintenance can be used to "force" the transition from being reactive to becoming proactive.

2. The predictive and preventive tasks identified during RCM analysis can be used to specify the rounds to be performed by operators in organizations focusing on implementing Total Productive Maintenance or Operator-Driven Reliability.

3. The conversion from the preponderance of reactive tasks to mostly proactive tasks can help increase the proportion of work that is well planned and tightly scheduled. (Studies have shown that well-planned and tightly-scheduled work can be done with as little as one-fourth of the amount of resources needed by unplanned and unscheduled work. So this effect can significantly reduce spending and resource demand.)

4. Fewer reliability failures will result in increased availability of productive resources. And increased availability will result in increased production capability from any asset.

5. RCM will help identify the kinds of chronic Failure Mechanisms and Failure Modes being experienced. This knowledge can lead to programs other than simple maintenance to improve reliability and availability performance.

If the objective is to sell the new program being compared in the assessment, the presenter will want to highlight all the possibilities. Some organizations will not harvest all the benefits, but that is their choice. They at least, need to understand what is possible and what others have been able to accomplish.

It is best to kick off the assessment with a presentation to as large an audience as possible to hear about the benefits. Although some members of the audience may be unable to help build enthusiasm, if the new process makes sense to them, at least they will not be negative. During this kick-off presentation, the audience should be provided with a comprehensive vision of the expected performance.

As an example, if a presentation was aimed at assessing an organization's current level of proactive maintenance, with the objective of introducing RCM-based maintenance, it would make sense to describe all the characteristics described above. Although some of the audience may be wagging their heads "no" and saying "It will never work here", others will be thinking of ways that the changes will make their lives easier.

How Good will Performance be, using the New Process?

The core of any assessment is a comparison between the current situation and some new or different way of doing things that will result in improved performance. To make this comparison, it is necessary to create a "vision" of the level of performance that should be expected when all the changes have been implemented. This sounds difficult and it is. But, it is not impossible. Taking the example of a complete RCM implementation as a basis for discussion, the list of benefits above included several characteristics, each of which can be converted into tangible indices and for those indices, measures of "excellence" can be identified.

As an example, take the characteristic of "proactivity". For the sake of this discussion, "proactivity" is further defined as a measure of the proportion of proactive work actually being accomplished. It would be possible to measure "proactivity" in several ways, so it is important to also define how it is to be measured. The following example shows how to define the measure of "proactivity" so that it can be used as a basis of comparison:

1. Identify the number of man-hours of Predictive Maintenance being scheduled.
2. Identify the number of man-hours of Preventive Maintenance being scheduled.
3. Add the Predictive and Preventive maintenance hours together to obtain the total amount of Proactive Maintenance.
4. Determine the total number of man-hours of maintenance work being scheduled. Be sure to compare

apples-to-apples by eliminating unproductive man-hours from both totals.

5. Calculate the percentage of Proactive Work by dividing proactive work by total work and multiplying by 100.

6. Perform a follow up verification of the amount of work that is actually being done. This step assumes the likelihood that some portion of the work being signed off as complete is actually being "paper-whipped".

The next step is to identify the standard of excellence that would be expected when performance has been improved by implementing RCM-based maintenance. For this example, it may be said that the objective is 80 percent. In other words, 80 percent of all work is proactive and only 20 percent is reactive.

Each and every area of improvement should be translated into a tangible characteristic and from there into a measurable index for which a measure of excellence can be identified. As with "proactivity" above, each of these characteristics needs to have a clear methodology for conducting the measurements, so that the comparison between the objective and the current performance can be meaningful.

How Bad is Current Performance?

Using the specific characteristics and measurement methodology for "good performance" that was introduced above, the same methodology must be used to assess those same characteristics in the current situation. Continuing with the example of "proactivity" as a basis for identifying the value of performing RCM and introducing RCM-based maintenance, it is necessary to quantify the current level of "proactivity" using the approach described above.

To accurately assess this characteristic, it will be necessary to create a report from the current CMMS (Computerized Maintenance Management System) file showing:

- Current Predictive Maintenance man-hours
- Current Preventive Maintenance man-hours
- Current Total Maintenance man-hours
- Actual non-productive man-hours

These numbers can be used to calculate the current actual performance. Beyond simply performing the calculations, it is necessary to perform a hands-on evaluation. Of the portion of proactive work being scheduled by the CMMS, what amount of the work is actually being done and done in a manner that will produce the "intended results". For instance, if a PdM task has been "simplified" so that it can be performed by the operator. Ask:

- Is the task actually being performed?
- Is the task being done in a manner that will identify a defect if one exists?
- If a defect is identified, is it being corrected?

A comprehensive assessment needs to provide a clear picture of all elements that affect the issue in question. If, for instance, the assessment identifies the lack of PM/PdM in the work order system, but does not identify the fact that workers ignore the schedule being created by the work order system, the assessment will be incomplete. The final recommendations will include only a part of the needed effort and resources and the finished product will not be effective.

No one is happy if they have invested scarce resources and effort into a program only to find that still more effort is needed to achieve the expected level of success.

Is there Sufficient Support to Drive Success?

Probably the most significant element in implementing a change to a new or better way of doing things is the support needed to drive it through to completion. There are a lot of good ideas in the world. Some of them will produce success at one company and will founder at another. There is a myriad of reasons for this difference, but the most common is the degree of management support.

Personally, the author has been involved in a number of situations where senior management has seen a need for change and, as a result, has initiated a program that should have produced the needed change. In some instances, the senior managers have remained engaged for the long haul or until success was accomplished. In others, their idea became just another in an endless series of "programs de jour" or "today's urgent program". In the latter situation, the participants quickly learned how to survive without committing themselves, or interrupting their current patterns. They have learned to smile and to repeat back the "take aways" that their supervisors expect them to learn from the meeting. Experts can parrot phrases while planning for their week-end activities.

A financial advisor once told the author, concerning how the investment business frequently works:

"When an individual makes an investment, the broker makes money and the investment house makes money and . . . Well, two out of three ain't bad."

The same is true of many improvement programs put forward by consultants:

1. The consultant hired to support the implementation makes money.
2. The people assigned to lead the program gain some visibility and therefore are typically rewarded and make money.
3. It is only the business and the people working in the "trenches" who continue to suffer.

An honest assessment will provide candid feedback regardless of the outcome. If a client is implementing a program only because it is politically correct, but the program lacks the support needed to drive it through to success, the client needs to be told so. Several positive results are possible:

1. If the current level of support is inadequate because there are too many irons currently in the fire, maybe senior managers will choose to proceed with the new program and stop the others that are competing for resources.
2. If the current level of support is inadequate because there are too many irons currently in the fire, maybe senior managers will choose to delay implementation until resources are sufficient to make it successful.
3. Maybe this opportunity will be lost but credibility will be maintained, leading to later opportunities.

The Barriers?

The next thing any assessment should identify is barriers or obstacles. A barrier or obstacle is an entity that is different from all the effort that is required to implement a change.

A barrier or obstacle is something that is intended to hinder progress. Clearly, all the work needed to accomplish an objective stands between the current situation and the desired situation. That work is not an obstacle or barrier because there is no intent to hinder. A barrier or obstacle is an issue or a problem that has to be changed if success is to be achieved. An example of a barrier is the "power structure" in the current situation. If certain individuals are able to exert undue influence because of the way things currently work, they will be unwilling to give up that influence if a new, more rational, approach is introduced. To overcome that kind of barrier, it will be necessary to help the power structure people see the advantage that the new approach holds for them.

In almost every situation in which change is required "change" itself is a barrier. In his book "Managing at the Speed of

Change", Daryl Conner describes the psychological barriers to "change" and shows how to aid individuals in overcoming those barriers. Most people prefer that their lives remain predictable and try to avoid surprises. Even good changes make their lives less predictable and cause concern. If the new plan for improvement has not taken into account the natural resistance to change, it is unlikely that real change will happen.

What are the Realistic Performance Expectations?

If anyone has a sleepless night and resorts to getting up and turning on the television, it is likely that they have watched some of the infomercials selling either weight-loss programs or exercise programs. Almost all the commercials present a newly-manufactured device that can be used to reshape one or another part of the body. There are always one or more individuals who have that particular part of their body highly developed, and they give testimony saying that their appearance is a result of the new device that is being sold.

Think for a moment, how long it takes a normal person to lose a few pounds, and how long and how much exercise it takes to make any difference in one's appearance, if ever. That timeline is inconsistent with the amount of time that the new device has been available. So, in reality, the appearance of the person giving testimony has nothing to do with the new device. The gadget may or may not provide some improvement to the individuals who purchase it, but the main objective of the infomercial is to sell products by creating unrealistic expectations.

Returning to the subject of performing an assessment and portraying realistic expectations after the improvement has been made. An earlier section discussed creating a vision of the level of performance that should be expected. Usually, that is the "best case" result and will come only after a great deal of dedicated effort. In a more realistic view, the results are the "best case" diminished by the following factors:

- At most, the organization will focus on this program for a period of one to two years. At the end of that time, other demands will distract attention away from that project. Long-term performance will depend on what elements have become a permanent part of the culture by the time the attention fades.
- Some portion of the effectiveness will be eliminated because of limitations in support.
- Some portion of the effectiveness will be eliminated because of the need to spend resources in dealing with barriers.
- Some portion of the effectiveness will never be achieved because of limitations in the availability of optimum resources.
- And so on.

In evaluating the ultimate effectiveness of any program, it is best to provide a realistic portrayal of the results that should ultimately be expected. The realism will do two things:

1. It will help management to determine the best place to invest limited resources.
2. It will help to identify those elements that are diluting the effectiveness of the program and of other programs.

The second benefit is the most important. If the answer keeps coming back that we just cannot make this program work because of Old Joe, sooner or later management will come to the realization that they may be better off without Old Joe.

What does it Cost?

As mentioned earlier, maybe the most important part of any assessment is determining the cost of all the activities needed to change from the current performance to the desired performance. This change is critical for two reasons:

1. Costs are easier to measure than results. As improvements are being made, everyone becomes excited and

happy about the results, even if they tend to fall short of the original vision. On the other hand, the bean counters never get excited about much of anything. The project is either on budget or it is not and the difference is easily measured.

2. Everything has a value in relationship to its cost or the required effort. Although the results may be attractive, they may not be affordable. Or they may not be affordable in comparison with other opportunities or currently-committed programs.

Simply said, ethical people do not overstate results or understate costs.

At the conclusion of any assessment, the client is owed all the information needed to make the right choice. If the assessment provides only a portion of the information and the client makes an improper choice, the assessment was faulty. The following sections are provided to help ensure that a complete and accurate recommendation is provided.

Does the Foundation Exist?

The author visited a chemical plant a number of years ago that had no equipment numbers assigned to any of their equipment items. When the operators wrote a work order, it would say something like, "Fix Old Blue" or "Change the oil in Old Shakey". Unfortunately, some of the people called the charge pump "Old Blue" and some referred to it as just "Blue". Also, it appeared that some of the employees must have been color-blind because the pump was not very well maintained.

The point is that a numbering system that supports a computerized filing and tracking system is one of the "foundational items" that a plant needs. The term foundational item indicates that if the characteristic does not exist, it will be necessary to build it before any other meaningful effort can begin.

Although the lack of an equipment numbering system may seem like an extreme example, there are many things that many of us take for granted but that may not exist in all situations. Another such item is an effective CMMS (Computerized Maintenance Management System). Most organizations have an effective CMMS but others do not, and still others have systems that are cumbersome and a drag on organizational effectiveness. In many plants, the CMMS and the associated work management processes have not been implemented in a manner that will support the envisioned mapping programs.

The author has seen instances where the maintenance staff have religiously printed out CMMS work schedules every day, only to have the crews discard them and work on what they prefer. In such a plant, the current structure and discipline is unlikely to be capable of supporting any advanced solutions.

What Infrastructure Changes are Required?

First, it is important to define what is meant by "infrastructure". For purposes of this discussion, the term infrastructure will be used in the broadest sense possible. Infrastructure can include features of the organization, facilities, information systems, etc. In some ways, infrastructure requirements may overlap with foundational requirements, but may be more subtle. The issue here is not getting tied up with definitions, the issue is to identify all things that need to be addressed or changed to achieve the desired objectives.

An example of a needed infrastructure change would be a situation where two or several individuals who need to communicate regularly (continuously) are seated in different buildings. It would be better to provide those people with adjacent offices and that might require some facility changes.

Another example is a situation in which the current CMMS is adequate but not readily available at the time and in the places where it is needed. The solution might be more terminals, or it might even be hand held devices that can be used as terminals.

The point is that a complete solution needed to achieve the desired objectives may entail some spending that requires "capital dollars" and those funds may come from different sources or be scarcer. However, even complex issues can be solved if identified early enough.

What will it cost to Install the New Process?

For the sake of this discussion, it is assumed that the improvement opportunity being pursued will require some changes in the business process. Earlier, the example of converting to a more proactive approach to performing maintenance was cited, and that example will be used to explain this element of cost.

If it is desired to become more proactive in the way mainte-nance is performed, the following steps must be taken:

- Effective predictive and preventive maintenance activi-ties must be identified.
- Those activities must be built into tasks that can be accomplished by one of the available resources.
- If special tools are needed, they must be obtained.
- If special materials are needed they must be set up in inventories.
- If special skills are needed, the necessary training must be provided.
- The tasks must be loaded into the CMMS and be scheduled at the proper times.
- Audits must be performed to see that the tasks are being done properly.
- Measures must be taken to ensure that tasks are being done correctly and are having the desired effects over the long term.
- Over time the lagging measures that needed to be

impacted must be checked to ensure that the
changes are having the business impact it was
hoped to achieve.
- If the program upgrade consists of a myriad of similar
tasks, all the activities described above must be performed for each and every proactive task that is to be
implemented

The list above might sound like a lot of detail, but it is a realistic assessment of the steps needed to change from being reactive to becoming proactive. At times, people like to portray such a
change as being like "turning a light switch". But few things work
that way. As with any of the other elements described herein for
process assessment, it is better to be thorough and accurate up
front, than look foolish later.

What is the Total Impact?

Adding together the expenditures on all the individual elements described above will result in the total cost. Although it is
important to arrive at the total cost, it is even more important to
highlight some of the other (non-cost) impacts.

- If the program will depend on the involvement of specific
key individuals for success, it is critical to highlight
that fact. Those people may be committed to other activities and the proposed program may lack the importance
to pull them away from previous commitments.
- If specific infrastructure changes are needed, the "owners" of the affected systems must buy into the changes.
- If organizational changes are needed to remove barriers,
the current supervisors need to support the changes.
Many people support changes that add to their own
authority but will often resist changes that give the
appearance of reducing their importance.

When the final presentation is made, a critical element in presenting the conclusions of an assessment with recommendations is to include all those individuals who need to support the changes. It is typically best to preview the presentation with the primary sponsor or the "big boss" before showing it to others. But in the end, it is best to avoid providing an excuse for individuals who oppose recommendations for personal reasons to have a valid reason for complaining. One further step at the beginning of the presentation of recommendations is to establish the expectation that silence during the presentation infers consent. In other words, if participants do not openly object (or at least reserve their support) their silence is to be interpreted as meaning they support the changes.

In some instances, specific elements of a proposal can cause concern and be "glossed over". At those times, the "owner" will say, "I didn't think it was a big issue. I thought we could change the details later." In those situations the programs often end up becoming diluted by compromises, with acceptance of second and third choices rather than the more difficult choice needed to achieve complete success.

ROLLING BOTH VALUE AND COST TOGETHER, WHAT IS THE ROI?

The best measure for comparing activities that consume attention, effort, and resources is Return on Investment (ROI). Although it is simpler to calculate ROI for a tangible project like a new plant or a modification project, any program that consumes resources and is intended to improve performance and produce a return should compete for the needed resources using that same standard of comparison.

Different organizations have different ways of calculating ROI. The secret is transforming program results into a form that can be analyzed using the specific ROI calculation method being used by the company concerned.

To keep this ROI calculation simple, both the costs and the anticipated returns must be converted into either quarterly or annual amounts (depending on how the specific form of analysis is set up to accept outlays and incomes).

The amounts being spent and the time at which they will be spent can be determined by applying each of the tasks to the appropriate period on a Gantt chart just as would be done for a more conventional project. In addition to identifying total spending and spending rate, this exercise will be useful in providing additional information. This same Gantt chart can be used to identify the rate and timing resources that will be needed and will provide a link to the schedule for estimating the timing of returns.

The amounts and timing of returns are a little more abstract than the outlays. The work accomplished during the first part of the assessment is useful in identifying the total amount of improvement that can be anticipated for each of the areas that were assessed. But that is only the beginning of determining the anticipated return. The value of each of the improvements needs to be "dollarized" or turned into tangible amounts that can be compared with the present value of amounts being spent. Keep in mind that most of the effort will result in spending of current-day dollars and resources. On the other hand, if the program is as successful as hoped, it will continue to accrue returns for many years to come. In addition, many of those returns will tend to build on themselves, providing even greater opportunities for future success. The value of those returns needs to be reduced to a present value using amortization tables.

For example, take a program that is aimed at becoming more proactive. Many RCM programs contain a scheme for analyzing the value of recommendations for a specific system or an entire plant. It is unlikely that all the recommendations will be implemented at once. More typically, they are implemented over time. Once implemented, the changes resulting from the program will produce a variety of returns:

- The new proactive tasks will reduce the number of failures.
- The new predictive and preventive tasks will extend the useful life of equipment because they eliminate deterioration.

In either form, there will be some delay before the returns begin to accrue. The proactive maintenance practices need to "take hold" before they can produce a cost-saving effect. In other words, the tasks need to start preventing deterioration or eliminating defects for some time before the return on the initial investment is felt. Deterioration is an impact that is experienced over time, and the value of preventing deterioration is a benefit that also takes place over time.

If proactive maintenance that increases the mean-time-between-failure of a device from six years to ten years has been installed, it will be necessary to wait to the end of the six year period to see the extended life. Any assessment that offers an ROI on equipment life more quickly than the end of the current life is likely to be an overstatement of the facts.

On the other hand, immediate returns are available from becoming "proactive". Life extension requires some time to experience, but the increased maintenance effectiveness that results from enhanced ability to plan and schedule work can be captured almost immediately. That benefit will depend on having systems in place that take advantage of the opportunity. For instance, if a move is made from performing reactive work to performing proactive work, detailed planning packages can be created for the proactive tasks. These plans would be far more detailed than the plans that could be made if the planner was working in a reactive mode, fixing things that were broken. The planner can also weave the shorter, better-defined, proactive tasks into tight schedules that waste little of the available resource. In addition, materials can be pre-staged and inventories can be reduced because the planner will know when proactive jobs will come due.

On the other hand, if detailed planning or tight resource scheduling is not done, the opportunity created by becoming proactive will remain unharvested.

The secret of calculating the return on any new program lies in the following steps.

- Identify areas of improvement.
- Identify ways to dollarize the value of the improvement.
- Identify the realistic timing of when the improvement will result in tangible returns.

Once the amounts and timing of all costs and all returns are known, apply them to the financial analysis model being used and calculate the ROI. When discussing the anticipated ROI with stakeholders, be sure to add significant detail so that they understand what efforts will be required to harvest the return. For example, if distasteful organizational changes, or workforce reductions, are needed to harvest the benefits, make that point clear from the very start. To achieve the desired return, the workforce will need to be reduced.

Chapter 6
Characteristics Evaluated During a Failure Mapping Assessment

Climate is what we expect, weather is what we get. **Mark Twain**

Referring to the quote from Mark Twain, we live in a temperate climate with an average temperature of 74-degrees and less than one-tenth of an inch of moisture each day. Also in this region, temperatures are sometimes less than 40-degrees below zero, more than twenty-four inches of snow can fall in a single day, and tornadoes can measure a half mile across. So, although the average may seem quite attractive, the extremes are not so welcome. That statement really defines the difference between climate and weather. To sustain conditions much closer to the typical conditions described as climate, it is necessary to live in a controlled environment.

Conceptually, the same issues are true of characteristics that impact maintenance effectiveness and reliability. In general, those characteristics may be good, but the average is made up of a wide variety of extremes. To moderate the extremes, all activities must function within a "controlled environment". A controlled environment here is a structured business process designed to ensure that critical characteristics exist and that they are managed in a consistent manner.

Previous chapters described Failure Mapping as a process that documents all the steps that exist in the life of a failure. The steps that are part of a failure map, in a sense, are steps in one of the business processes of a plant. While creation of a latent cause or a physical cause or a human cause is not something done as an intentional part of a process, each and every organization makes conscious or unconscious choices at each of those levels of cause.

By way of example, consider those organizations that have installed highly-effective safety programs. Those bodies have made a conscious decision that worker injuries are unacceptable. Most of them have also made the conscious choice that unsafe acts and unsafe conditions are unacceptable as well.

Has your organization made the conscious choice that all reliability failures are unacceptable? No, well then has the firm chosen to believe that some failures are acceptable? If so, they have decided to accommodate a business process that allows some amount of failures, and in doing so, they are accepting latent causes, human causes, and physical causes, as a normal part of doing business.

The point is that tolerance to some failures suggests that a failure process is in place and functioning. The next question should be, "Is it better to have a failure process that is out of control or one that is under control?" If the failure process is to be managed, Failure Mapping provides at least one realistic alternative.

The objective of installing a FM process is not to simply have another business process to manage. The objective is to install a consistent process (similar to the controlled environment), that will ensure that a number of the desired outcomes will be achieved.

In assessing the need for FM, or determining the value that the FM process will provide, it will be necessary to investigate the characteristics that FM will provide. If the current systems fully provide those characteristics, FM will add little additional value.

On the other hand, if the present organization or systems do not provide the desired characteristics, significant value can be captured by implementing Failure Mapping.

The following is a list of characteristics that will be introduced or significantly enhanced with Failure Mapping. The amount of improvement or "gap" from the current state will determine the ultimate value.

CAUSE IDENTIFICATION AND ELIMINATION

The popularity of Root Cause Analysis has expanded dramatically in the recent past, but a myriad of failure causes still seem to be embedded in many organizations and practices. One might ask how the current form of Root Cause Analysis (RCA) is performing, compared with what is really needed. Ask the following questions:

- How deeply does RCA dig? Are there issues that are off limits? If so, why do they remain off limits?
- How frequently is RCA used? Is it used on all failures or just a small portion of the really significant failures?
- How many individuals have been specifically identified as the "human cause" for a failure? Or does the organization make it a practice to avoid naming names?
- In the form of RCA in use, is the naming of a "Human Cause" so infrequent that it carries with it a significant stigma for the person or persons involved?
- In the present form of RCA, is the naming of a "Latent or Systemic Cause" so infrequent that it carries with it a significant stigma for the manager of the organization or systems involved?

If the RCA process is used in a manner that fails to identify the behaviors of specific individuals that produce the physical causes, they will never change those behaviors. Similarly, if the RCA process is used in a manner that fails to identify the weaknesses or traps that exist in the organization or systems that allow the Human Causes to occur, they will never be corrected.

The point is that RCA has to be applied in a manner and with sufficient frequency to completely eliminate the stigma associated with "cause". Cause is not fault. Every person makes mistakes. The only person who doesn't make mistakes is the person who doesn't do anything. The same is true of organizations and systems. The existence of a trap is no surprise. The unwillingness to identify or admit to weaknesses is the problem.

When assessing the value of implementing FM, one of the first characteristics to evaluate is the ability of an organization to identify and shed the behaviors and weaknesses that cause failures. If the processes focus on only Physical Causes and then on only a small portion of them, there is a significant opportunity to reduce more physical causes, and all human and latent causes.

FAILURE MECHANISM IDENTIFICATION AND ELIMINATION

For the purposes of this discussion comments will be limited to Failure Mechanisms affecting mechanical systems only. These mechanisms are:

- Corrosion
- Erosion
- Fatigue
- Overload

There are analogous Failure Mechanisms that affect electrical systems, but the discussion can be made more concise by focusing on the smaller group. The reader is left with the mental exercise of applying this discussion to electrical failure mechanisms.

Depending on the kind of organization being assessed, inherent understanding of Failure Mechanisms and deterioration can vary widely. Some organizations have large "inspection" organizations that dutifully monitor mechanical systems for deterioration that results from various forms of corrosion. Formalized processes that monitor the metal thickness of key components at specific points and at specific intervals are in place, so that the rate of deterioration is monitored and the time of failure or renewal is determined and forecast.

Other companies have no such functions, and they focus their limited resources on the specific issues that are immediately associated with operating the business. When a failure occurs and the cause is uniform corrosion, these firms are likely to guess the cause correctly and take some action that will delay deterioration in the future. When a failure is the result of some more abstract Failure Mechanism like erosion due to cavitation (for which there is little remaining evidence after the failure has occurred), or fatigue, the failure is attributed to misfortune or "bad luck".

When assessing the value that FM can have on the organization, the two examples described above would show wide differences in the amount of improvement that is available.

In the first example, the application of FM and the creation of a mature FM culture would tend to expand knowledge of Failure Mechanisms to a much broader population. If more members of the organization understand how Failure Mechanisms like corrosion, erosion and fatigue work, these people are more likely to recognize the symptoms while the failure mechanisms are working but before they have caused enough deterioration to produce a failure-causing defect.

Imagine a situation where everyone in the organization can recognize that excessive vibration could be causing fatigue. That vibration could be arrested before deterioration and failure can occur.

In comparison with the first instance above, FM will make "part-time inspectors" out of the entire staff. Rather than having a few inspectors looking for a few Failure Mechanisms, there would be a lot of eyes and ears looking and listening for all Failure Mechanisms.

In comparison with the second instance, FM will create the technical assurance needed to ensure reliability and equipment integrity without the burden of a group of individuals (inspectors) who are focusing their entire effort on activities completely outside the sphere of revenue-generating activities. There is a much greater opportunity in the second situation, but there is also a different kind of opportunity. If an "inspection department" does not currently exist, FM provides the opportunity to have all the advantages of the benefits of an inspection department without the added cost of maintaining one.

PRECISION REPAIRS

Again, think about how repairs are completed by the current organization.

- Are the repairs permanent?
- Do the repairs make the item "as good as new" or "as good as old"?
- Do the repairs address the real defect of simply "restore operation?"
- What is the typical life after a repair?
- How long do repairs require? Is all the downtime used effectively or are there several "false starts" before finding the real problem?

When assessing the current practices of an organization in making repairs, it is important to review the few instances each day that seem to require more attention than seems appropriate. Although these instances may be the exception rather than the rule, it is these few that make life difficult for everyone. They consume more resources than necessary. They take more time than necessary. And they divert the attention of limited staff away from more strategic issues.

One of the most significant features of FM is that it points at the most likely Failure Mode and then sequentially points at each other Failure Mode based on statistical likelihood. By this sequence, FM prevents your organization from getting tied up in those complex problems that have some history. It is only those of the Failure Modes that are brand new that require escalation, and then not all of them. Only those brand new Failure Modes that are not identified while historic Failure Modes are sequentially investigated will require escalation.

In addition, because repairs are aimed at highly-likely Failure Modes, when the troubleshooter finds the current Failure Mode, it is fairly certain that a permanent repair will ensue. With FM, the statistics will be in favor of the company.

The assessment of value resulting from FM will be based on how quickly the right answer will be found. That characteristic has two distinct parts:

1. How quickly is the solution arrived at?

2. How often is the repair permanent? In other words, how often is there a repeat failure in a short amount of time?

Both of these characteristics can be dollarized based on equipment unavailability and the repair costs resulting from that unavailability.

DIRECTED TROUBLESHOOTING

How does the organization conduct troubleshooting? Or does it conduct formal troubleshooting at all?

All organizations conduct troubleshooting in one way or another, and most organizations conduct it in a fairly informal manner. Unless specific attention is given to the way malfunctions are reported, it is likely that the person assigned to find the problem and make the repair only knows there is a "problem". That person is assigned to "fix the problem", not find the defect. If the problem can be fixed without searching for or finding the defect, all the better. That situation will typically result in a much quicker and cheaper fix than if a real live defect has to be dealt with. The problem is, that approach does not eliminate the defect or produce a permanent solution.

Taking the first step toward improving this activity will entail creating the function of "troubleshooting" within the organization. In a normal situation when it is decided that an effective troubleshooting activity is needed, a person will be selected who has a clear understanding of how things work (understands the physical systems), and knows how they deteriorate and break. The job of the troubleshooter is to keep looking until a defect is found that has been translated into a Failure Mode. The Failure Mode is a specific component portraying a condition that explains the Malfunction Report (impaired function and behavior) that was recorded.

Depending on the organization, the troubleshooter may or may not make repairs. In a high-volume operation, good troubleshooters are typically a scarce resource, so less-skilled individuals are assigned to perform repairs. The troubleshooter can then go on to the next problem.

The second step toward improving performance will be to apply the elements that make the troubleshooter more effective.

- First, if the report of the problem begins with a "Malfunction Report", there is a much more clear understanding of the starting point for the repair.
- Second, if there is a record of specifically which Failure Modes (component – condition) have caused the specifically-identified Malfunction Report in the past, the trou

bleshooter will know the several alternatives that are possible.

- Finally, if records of the frequency of Failure Modes have been maintained, the statistical likelihood of each alternative will be known. As a result, the troubleshooter will know the best place to begin the search for the defect, where to go from there, and so on.

When assessing troubleshooting, or more specifically, directed troubleshooting, the objective is to determine how closely the organization being assessed resembles either of the various levels described above. In one instance, no formal troubleshooting is done, it is simply smeared in with the repair. At the other extreme, the troubleshooting is provided with all the tools and information needed to identify the specific defect quickly.

Rather than looking at individuals and their job titles, start by determining how quickly the person assigned to a repair can identify the defect. This issue combines two measures:

1. The speed of finding the defect.
2. The accuracy of finding the real problem.

The first characteristic can be measured with a clock. The second characteristic is measured by the number of repeat failures that result from failure to identify and eliminate the defect. In some circumstances the current system is fast but inaccurate. In others, the system is dreadfully slow but thorough. The objective should be both speed and accuracy.

STRUCTURED DATA-DRIVEN DIAGNOSTICS

As with troubleshooting, it is useful in discussing Structured Data-Driven Diagnostics to ask how the organization currently performs diagnosis. Many people might ask specifically what is meant by diagnosis. For the purposes of this discussion, diagnosis is the hands-off step that occurs before the troubleshooter is ever dispatched to the site of the problem. This step makes use of the following information to make the first "guess" at what is wrong:

- Malfunction Report
- Remote Computer Control System Downloads
- Maintenance and Repair History
- History of Failures
- Failure Maps
- Funneling Recommendations
- List of Likely Failure Modes
- Statistical Likelihood of Each Possible failure Mode

As might have been guessed, the mine has here been "salted" with tools that are available only in a mature FM environment. With the amount of information described in the list above, the diagnosis being provided is far more than a "guess", and is likely to be an accurate call.

In assessing the value that FM will add to the current diagnostic process, a good beginning is to sample the first individuals being assigned to respond to a failure in the current organization.

- What do they know about the problem?
- Do they know where to begin looking for the failure-causing defect?
- If the first recommendation fails, do they know what to try next?

In fact the real question is, do these people have any idea what the problem is? It is also valuable to follow up after the job was completed. Did the first person assigned repair the problem, and did he even have the correct skills? Also, how many trips were needed from the shop to the worksite to get the right tools and parts?

In a mature FM environment, the troubleshooter with the correct skills is likely to be dispatched first, and the right craftspeople, with the right tools and parts, will follow after the troubleshooter has quickly identified the defect.

In a mature FM environment, the decreased cost of each typical repair job can be "dollarized" (comparing the approach with FM to the approach without FM), and the typical number of repair jobs in any period can be summed to identify the total value in structured diagnostics.

FUNNELING

Many organizations may not even understand what this concept consists of. Here is an example:

> Let's assume that one of the primary functions of a system is to produce electric power. Let's further assume that this system consists of a prime mover (a diesel engine) and a generator. Let's continue by assuming that a malfunction occurs and the function lost is "power generation".
>
> In this situation, the individual reporting the malfunction only knows that the power generation function has been lost. The best description of the behavior is that power generation has dropped to zero.
>
> In the past there have been both electrical (generator) and mechanical (engine) Failure Modes that have resulted in that apparent behavior. To speed up the response, it would help to know if the ultimate problem is electrical or if it is mechanical. The present staff of troubleshooters tends to have either electrical skills or mechanical skills, so sending the wrong person is likely to delay resolution.
>
> In this situation funneling is the activity of gathering additional information to assist the process of directing the troubleshooting to the most likely cause.
>
> In this example, let's assume that the diagnostician performs the funneling step by determining if additional symptoms exist. For instance, maybe the diesel engine has been blowing clouds of black smoke prior to the failure. Or the operating records over the last few days prior to the failure show that the coolant temperature has been trending up close to the trip point.
>
> With this added information, the diagnostician could reduce the likely alternatives to the mechanical alternatives only and dispatch the troubleshooter who has the greater mechanical skills.

When assessing funneling, one could begin by simply determining whether an organization understands the concept. If the concept is clear, then it might be possible to review past work orders from several sources. In those events, who was assigned? Was it the resource that was most readily available? How often has another person had to be assigned after the first person reported back with the comment, "This isn't in my area".

Time is money and mechanisms that save time can be dollarized into value associated with this element of FM.

Malfunction Reporting

A responder is a person who takes action when a problem is reported. Take a moment to review how problems are currently being reported in the organization. In the excitement of responding to a problem, the responder often provides a colorful description of everything that occurred surrounding the problem, but not the problem itself.

A responder really needs to know which function has been affected and how that function is currently behaving. The people who currently report breakdowns need to know what functions are being provided by the equipment they operate or oversee. They generally understand what a system does, but not in a well-defined manner that will aid them in directing a response in the event of a breakdown.

For instance, in several reports concerning the failure of an air horn system in which the air to the horn was controlled by an electro-magnetic valve. Over time, this valve had developed a bad reputation. A number of the reports said:

- The magnetic valve was stuck
- The magnetic valve was broken
- The magnetic valve was froze

Unless the individuals making the reports had X-ray vision, it was impossible for them to make any valid report that involved the condition of the magnetic valve. Alternatively, it would have been realistically possible to report:

- Horn alarm – No Sound
- Horn alarm – Intermittent Sound
- Horn alarm – Weak sound

With "no Sound", it would be possible to supplement the report with information concerning operability "above freezing temperature" as compared to "below freezing conditions". Each of the initial reports described above would provide a meaningful starting point for troubleshooting, but the added information concerning functionality above and below freezing would tell the troubleshooter that the system was working but had perhaps have collected some moisture from the air, and that the moisture was causing a frozen condition when temperatures dropped.

Even in a non-FM environment, it is very helpful, to provide the discipline of reporting problems using Malfunction Reports. The ultimate value will not be as great as in an FM environment but it will still be helpful. In an FM environment, there will be a requirement to determine Malfunction Report alternatives that conform to the all the identified relationships that have been pre-identified in the FM database. Identification of random Malfunction Reports (Function-Behavior) that do not fit the established patterns would allow inexactness and inefficiency will not be allowed by the system.

In assessing this way to improve responses, and the value that can be achieved from implementing FM, it is worthwhile investigating current problem reports to see how well they identify problems. If there is a "loose fit" between current problem descriptions and the possible Failure modes, the amount of time needed to respond to failures will be increased, costing additional resources and money.

DEFECT RECOGNITION

As with improving the recognition of Failure Mechanisms at work, there is an opportunity to identify defects before they cause failures. Many defects create only one of the several conditions needed to trigger a failure. But there is time between formation of the defect and when other conditions are right for the failure. In this situation there is an opportunity to find the defect and avoid the failure.

In a mature FM environment, a significantly larger proportion of the organization's population than usual will know:

- What a defect is
- What defects have occurred in the past
- Where specific kinds of defects have occurred
- How to report a defect

In assessing the opportunity for improved Defect Recognition as a result of installing FM, some investigation is needed. For instance, it is useful to record how many times any fairly-apparent defects have gone unidentified until after a failure has occurred.

Part of the value of installing FM is forcing the organization to become more proactive and prevent more failures by identifying failure-causing defects before the failure can occur.

SUMMARIZING

FM provides a structured system for organizing the historic relationship between Malfunction Reports and Failure Modes. In doing so, it provides a basis upon which:

- Failure Mechanisms can be identified
- Root Causes can be identified
- Reliability can be improved

- Maintenance response can be improved
- Failures and the impact of failures can be reduced

The value of each of these opportunities can be dollarized, though it requires some work and record keeping. Comparing the total value of future opportunities with the value currently being harvested will determine the value of implementing the FM process.

Chapter 7
Assembling a
Failure Mapping Process

When it comes to getting things done, we need fewer architects and more bricklayers. ***Colleen C. Barrett***

If it is believed, when the assessment is completed, that the organization would benefit from installing a FM process; this chapter will tell how to do so. The quotation at the beginning of this chapter provides some useful insights into how the FM process is to be installed. The participants need to be "thinking bricklayers" who invest their efforts as well as their minds. Many of the activities need to be accomplished routinely, but they are all "thinking and learning" functions that perform today's work as efficiently and effectively as possible, but are being done while thinking about how to improve tomorrow's work.

Implementation of the FM process follows a number of sequential steps. These steps can and are likely to be done differently by different organizations, and the sections below provide a general description of each step.

CREATE THE FAILURE MAPPING TEAM

The implementation of the FM process begins with deployment of resources. The FM process will ultimately reduce costs and eliminate a number of the current resources needed for mainte-

nance, but it will begin with the
application of resources to complete
both start-up and on-going tasks.

There are five key functions
that are accomplished by members
of the FM Team as a part of the FM
process. They are:

- Failure Mapping
- Diagnosis
- Troubleshooting
- Failure Analysis
- Root Cause Analysis

These functions may be within the compass of current positions within your organization or they may be entirely new positions. The important issue is not who, but what, when, and how.
First of all, the functions need to be done. Talking about doing
something and trying to do something don't count. The actions
need to be done and done in a timely manner. They need to be
accomplished in a complete and accurate manner. The individuals
assigned to perform these functions cannot do them as a part-time
"hobby". The functions must be viewed as the job upon which the
performance and compensation of the individuals is measured.

The points above are stressed because the author has
observed a lot of the individuals assigned to work on specific functions, in addition to their day-to-day tasks, act as though they were
separate and distinct from the day-to-day workings of the maintenance process. In a FM environment, these individuals become a
part of the fabric of day-to-day maintenance. They add to the effectiveness of the maintenance process by increasing the information
that is available from the failure map database.

To avoid any confusion it is necessary to emphasize the
point that the Failure Mapping positions are not long-term additions

to an organization. These activities are already being done by every organization, only in a less-structured and less-organized manner, and typically scattered among a wide variety of individuals who have limited training and conscious experience. For instance, diagnostics and troubleshooting are accomplished by someone in response to every failure, and are typically done without ready access to all the critical information, and as part of the repair process. By concentrating diagnostics into the hands of one (or a small number) of individuals who have access to the organized history of failures in the organization, response will become much quicker, more accurate, and more effective. It is also possible that the whole organization will learn from all the extra experiences.

Although some extra effort will be needed to get over the "hump" of collecting information and organizing the failure maps into an easily accessible database, the individuals concerned will soon blend back into the day-to-day organization, performing their functions in a new, more-efficient and effective manner. Beyond that point, improved responses, improved effectiveness, and improved reliability, will ultimately reduce the resource requirements and maintenance costs.

It is suggested that the process be started by selecting five very capable individuals (who possess the attributes that will be described below) and telling them that Failure Mapping is their highest priority until the process is working smoothly. Each of these five should be given one of the roles based on his particular talents.

The group will need a leader. The leader can be either a separate Project Manager or one of the five. None of the positions provides what ultimately will appear to be a natural fit for the leadership role.

For instance, although the individual who, primarily, will be assigned to assemble failure maps is likely to have the broadest global view of the process, he will be the busiest during the time when attention to project management is also most needed. The other positions should be assigned to individuals with good analytical or mechanical skills, but not with particularly strong project management skills.

For the moment, let's assume that a separate Project Manager and five other individuals have been assigned to implement FM. Let's assume that the team selected has the following attributes:

- The Project Manager has strong project management skills and the ability to maintain a global view of all activities and objectives.
- The Failure Mapper has a good understanding of all the systems, their functions, the behaviors and symptoms exhibited upon failure, the Failure Modes, and the information systems used to assist in diagnosing and troubleshooting problems. This is a person who can back away from being directly involved in the day-to-day activities in the field, to focus on longer-term tasks and objectives.
- The Diagnostician has many of the same skills as the Failure Mapper but is a person more interested in the hectic pace of day-to-day activities, though from an office setting. He should be a person who seems able to recognize patterns and their relationships to downstream events.
- The Troubleshooter is a person with excellent manual skills, along with ability to recognize defects and under stand Failure Mechanisms, and how they cause defects. The Troubleshooter is the one person of the group who needs to keep moving 90 miles an hour and enjoying every minute of it. His role is one that is a link between the office and the field. He needs to be able to accept the intelligence being provided by the Diagnostician. He also needs to be credible with field craftsmen so that they accept his directions on repairs.
- The Failure Analyst is best suited to an engineer. This individual must be educated on the techniques needed to identify and recognize all Failure Mechanisms. He

must be able to do so without turning every incident into a Masters' thesis. He will need to identify the Failure Mechanisms, record his results, and get on to the next assignment quickly. He also needs to have the courage to push back when the troubleshooter has not identified the actual Failure Mode, and to work closely with the Root Cause Analyst in identifying likely causes.

- The Root Cause Analyst is an individual who will accept only facts. He must be trained on how to perform Root Cause Analysis (RCA) and how to facilitate the participation of others in performing RCA. Once again, this needs to be a form of RCA that happens very quickly and must not be allowed to turn into a project. The analyst needs the patience to gather all the necessary facts but also the impatience needed to recognize when enough is enough. He needs the credibility required to-gather information from factory floor workers and the maturity needed to work with managers and senior managers. He needs good speaking and writing skills and the ability to identify human and latent causes without being offensive or confrontational. He also needs to know how to pick his battles identifying the places to invest his efforts that will produce the most profound and far-reaching effects.

CLEARLY DESCRIBE THE OBJECTIVES

Once the FM Team has been formed, the first thing to do is to be sure that the team (and you) clearly understands all the objectives of the FM initiative. All too frequently, groups are formed and initiatives undertaken with the assumption that people will "just naturally" know why they are there. Leaders believe that the reasons are apparent to everyone, and the objectives must be equally clear. Individuals on the shop-floor do not share the same information as their leaders, and do not see things the same way. As a result, if management wishes to be certain that the FM Team understands the desired results, they must be told what is wanted and made to repeat it back to the manager.

Here is a review of the objectives of the FM process. These objectives occur at several levels and some of the deeper objectives are an outgrowth of those objectives that are more apparent.

1. The first objective is to improve reliability.
2. The second objective is to improve maintenance effectiveness and efficiency.
3. Reliability will be improved by:
    a. Performing permanent repairs of defects.
    b. Identifying Failure Mechanisms and eliminating them.
    c. Identifying Root Causes and eliminating them.
4. Maintenance effectiveness and efficiency will be improved by:
    a. Providing more effective and accurate diagnosis of problems.
    b. Providing faster and more accurate troubleshooting.
5. Accomplishment of the objectives described above will be facilitated by:
    a. Identifying the relationship between Malfunction Reports and Failure Modes
    b. Tracking the statistical likelihood of each Failure Mode.
6. Achieving the objectives above further depend on:
    a. Properly identifying all Malfunction Reports in a form that can be tracked and analyzed.
    b. Properly identifying all Failure Modes in a form that can be tracked and analyzed.
    c. Identifying the symptoms or other information that points the way from a specific Malfunction Report to one or more Failure Modes that could have caused the failure.
    d. Identification of the Failure Mechanism that resulted in each Failure Mode.
    e. Identification of the Root Causes that allowed the Failure Mechanism to become active.

7. Organization of all the above information and activities depends on implementation and maintenance of a system in which everyone does his part.

8. Once the system is in place, it will be possible to achieve much broader benefits by engaging all members of the organization in the knowledge and activities it involves.

EDUCATE THE FM TEAM

Once the FM Team understands their objectives, they will be anxious to get started. Experience has shown that if there are five members of the team, they will want to head off in five different directions. One of the first and most important issues to address is to keep all the team's efforts focused in a single direction. There are subtle differences in how a Malfunction Report or a Failure Mode can be defined that will cause significant confusion as the process grows. A simple issue like the words selected to describe functions, behaviors, components, or component conditions, can make a big difference as additional people begin to be involved in using the process.

As more people become involved in the FM process, it will become increasingly difficult to keep the terminology used in the process under control. People like to use their own creative names for things, and if the appropriate level of structure and discipline does not already exist, they will do so. Use the initial educational setting to identify some of the terms (like functions, behaviors, associated with each function or component condition) so that the group will learn their importance to the process and to failure mapping, and will be committed to their proper usage.

The education of the FM Team should include the following elements:
- The Path to Failure
- Elements of the FM Process
- The Roles of Each of the FM Team Members

All the basic information needed to develop training packages for each of those subjects is provided in this text.

CREATE THE INFRASTRUCTURE

Probably the most significant challenge in implementing the FM process is setting up an efficient infrastructure to support the process. Rather than simply providing a description of the infrastructure, it is better to describe how the systems will be used. It is left to the reader's creativity to identify how best to assemble systems that will provide that functionality.

The first element of the infrastructure is the database in which all failure maps will be stored. Keeping in mind that each failure map is a record of a specific failure, the database must contain fields that are designed to store all the pertinent facts describing the life-cycle of a failure.

Some of the specific needs of Failure Mapping are well supported by computerized database programs like Excel or, better yet, Access.

The following is a list of the information that must be added to the database with each failure incident. The database should provide fields for elements like Failure Mechanism and the three levels of Cause even though they will not be added immediately at the time of the failure, and sometimes may never be added.

Malfunction Report
- Date of Failure
- Equipment Identification Number
- Narrative Description of Failure
- Affected Function
- Description of Behavior

Close Out Information
- Failed Component

- Component Condition

Failure Analysis Information
- Failure Mechanism

Root Cause Analysis Information
- Physical Cause
- Human Cause
- Latent or Systemic Cause

In addition to the above information which is added to the database at the time of failure, when closing out the repair, or when performing the investigation, there is some additional information that should be added during failure mapping. This data will be used when the diagnostician is attempting to narrow down the alternative Failure Modes, or when he is performing triage, attempting to identify the best way to respond to the failure. That information includes the following items that should be added to the database in fields between the Malfunction Report and the Close Out fields:

Funneling Linkages
- Physical Symptoms
- Earlier Behavior
- Computer Alerts
- Systems where Failures can Cause this Malfunction Report

Subsystems or Components where Failures can Cause this Malfunction Report

Specific fields including Function, Behavior, System, Subsystem, Component, Component Condition and Failure Mechanism must be restricted to a small list of realistic choices using look-up tables. If responders are allowed the flexibility of using words not in the lists, the value of the database will be jeopardized.

- The Functions are the same functions used when performing Reliability Centered Maintenance analysis.
- The Behaviors can include a wide array of actions, just as long as the list represents only realistic choices and does not include duplicates or overlapping terms.
- Systems, Subsystems and Components frequently have a tendency to overlap. It is necessary to be certain that there is some value in creating the distinction and that the "Component" is the lowest level with which the organization typically deals. For instance, if the organization only changes out complete microprocessors, that will be the component level. On the other hand, if defective circuit boards can be identified, and they are commonly exchanged, that will be the component level.
- Component Conditions can be any number of descriptive choices. Again the challenge is to limit the number of items to those needed to be descriptive without overlap or duplication. The following are examples of terms that have been used to represent the same meaning as "over heating":

- Burnt
- Cooked
- Fried
- Burned up
- Melted
- Etc.

Failure Mechanisms are limited to the following:

- Mechanical
    - Corrosion
    - Erosion
    - Fatigue
    - Overload

- Electrical
  - Overload – Supply Transient
  - Overload – Load Stall
  - Electrical Equivalent of Fatigue
  - Insulation Breakdown due to Heat
  - Insulation Breakdown due to Chemical Attack
  - Mechanical Abrasion
  - Mechanical Loosening

In addition to the database used to collect all relevant information concerning failures, there must be a system for quickly accessing the information needed to perform the tasks assigned to the FM Team. These tasks are:

- Diagnosis
- Funneling from Diagnosis to Directed Troubleshooting
- Troubleshooting

The reporting system can either be a spreadsheet that the user can step through, or an automated database that has the ability to drill down as increasingly specific details are added.

The automated database might work as follows:

- Select – Function
- See – All appropriate Behaviors

Select – Appropriate Behavior
    See – A list of all possible questions that should be asked, or data that should be checked to funnel down to the next level. (Depending on complexity and the number of levels included, this may include several steps.) In the end, this should portray all the possible Failure Modes in rank order.

Select – Most Likely Failure Mode
    See – Additional information concerning how and where

repair work can best be done. If there is a quick-fix technique available (like recycling the control computer or power supply), the system will provide the appropriate instructions for the operator. Otherwise, instructions for the troubleshooter will be issued concerning where to begin his investigation and, if unsuccessful, where to go from there.

Once the current repair has been completed, the details describing it will be added to the FM database and to the statistics that determine the direction of future responses. Then the Failure Analyst and Root Cause Analyst will perform their investigations and add details of their results. Individuals who later perform Reliability Centered Maintenance analysis, Pareto analysis, or other forms of analysis aimed at addressing opportunities for improvement, can use the information contained in this database.

Once the FM database is set up, two steps remain to establish the infrastructure.

The first step is to identify the terms that will be allowed to be used in filing Malfunction Reports and those that will be allowed to be used in describing Failure Modes. Malfunction Reports are a combination of system functions and behaviors (or misbehaviors). Failure Modes are a combination of components and conditions. Caveats to be applied in creating these terms have been described earlier. It was suggested earlier that some work in this area be done during FM team training. It should be anticipated that creation of a complete list will require quite a lot of time.

The second of the two steps needed to complete the system infrastructure is to begin populating the database. Hopefully, past maintenance records will provide enough information to get started with this activity. The initial failure reports should be adequate to determine which system function was impaired and how it was behaving when the failure occurred. In addition, maintenance records, warranty records, or purchasing parts consumption records should describe which components have been failing. Maintenance records should also provide some detail concerning the condition of any failed components.

Information in the Malfunction Report (Function-Behavior) and the Failure Mode (Component-Condition) will permit several critical fields in the FM record to be completed. Using this starting point, the Failure Mapper can fill in the remaining fields such as symptoms, or computer alerts that will point the way from a Malfunction Report to a specific (or at least a smaller number of) Failure Mode(s). Using records of past failures provides a starting point for the database. A significant portion of the failure maps will be created during the first few months or years that the FM process is going through the maturation process.

DESCRIBE HOW THE PROCESS WILL WORK

The next step in creating the FM installation is to create a description of how the process will flow. The system will necessarily function somewhat differently while it is in the process of maturing than after it has matured.

While the FM installation is the process of maturing, the FM Team will have to depend on a limited number of maps and the limited statistics available from historic records. After the FM process is mature, each new event will have less impact on the database. New failures will add increasingly fewer new maps and will primarily reinforce the statistics that have already taken shape in the database.

Once the FM system is mature, the day-to-day functioning of the process will follow the pattern:

1. Individuals reporting a failure will select the appropriate function from a look-up table. The behaviors that are appropriate for that function will have been used to populate the behavior look-up table, and the person filing the report will select the most appropriate behavior from the table.

2. The Diagnostician will query the database to find all possible alternative Failure Modes.

3. If there are many possible paths that the repair can take, the Diagnostician will check to see what additional information is needed to complete funneling.

4. If funneling is needed, the Diagnostician will follow the steps described in the database. In some instances he might ask the operator more questions concerning symptoms or alerts being provided by control system screens. In other situations he may need to check maintenance records or operating records that are available from on-line systems.

5. When all possible funneling is complete, the Diagnostician will create a rank-ordered list of likely Failure Modes. (A well-designed database should do this automatically.)

6. The Diagnostician will compare the list of possible Failure Modes and who/where they can be repaired with the resources at his disposal. He will select the resource that is most likely to be able to dispose of the problem most quickly and efficiently.

7. There are several possible ways in which the Diagnostician may elect to pursue the solution:
     a. If the best resource for handling the problem is the system operator, the Diagnostician will call that person and provide instructions for the quick fix. He will wait to see if the quick fix worked and if so, he will record the close out. If not, he will go on to the next alternative.

     b. The Diagnostician next assigns the task to the resource most likely to be able to deal with the next most likely Failure Mode. Usually that will be a Troubleshooter. In that situation he will dispatch the

Troubleshooter with the remaining list of most-likely Failure Modes.

c. Sometimes, the failure is a wreck, and the proper first response will usually involve invasive work. The Diagnostician will then dispatch a repair crew directly.

8. If the operator is unable to correct the problem and the next most likely cure involves a Troubleshooter, the Troubleshooter will go to the site of the system and begin whatever steps are needed to isolate the cause of the defect. He will proceed sequentially, working down the list provided by the Diagnostician.

9. When the Troubleshooter identifies the problem, he will make a record of the damaged component and its condition, using terms provided in the look-up tables in the database.

10. Depending on the extent of the problem and the characteristics of the organization, the Troubleshooter will either make a repair or write a work order to have the repair completed by others.

11. When the work is complete and the FM data has been input, the defective component will be stored away, along with any evidence needed to point the way to the Failure Mechanism responsible.

12. The Failure Analyst will identify the Failure Mechanism and will pass along any evidence that may help identify the cause to the Root Cause Analyst. He will also work with the appropriate maintenance or engineering resources to see that the Failure Mechanism is eliminated. He will also work with maintenance planners to install Predictive or Preventive Maintenance procedures that will intervene before the next failure is allowed to occur.

13. The Root Cause Analyst will use the data provided by the Failure Analyst to begin investigating the three levels of Root Cause. When this investigation is complete, he will prepare a very brief report that will be issued to the individuals who are responsible for taking corrective action.

What is the Schedule?

It is always helpful to create a Gantt chart or physical schedule for activities and milestones. There are a few evident milestones at the very start, like producing the first round of failure maps that are based entirely on historic data. But after that, the process will get into a "rhythm" of generating failure maps internally and there will be fewer milestones upon which to measure progress.

Although there are fewer indicators, it is still possible to create some measures that can be used to drive progress. For instance, the time at which the Failure Mapper has begun creating maps using data from new events is a significant milestone. Another milestone is when the Diagnostician and Troubleshooter become actively engaged in the current maintenance process. A third milestone is when the Failure Analyst first identifies a Failure Mechanism and the Root Cause Analyst identifies the first Root Causes. There should be target dates for when the FM Team begins to have traction as shown by problems being solved (Failure Mechanisms eliminated and Causes eliminated), and the amounts of time spent finding defects and correcting problems is getting less.

Another aspect of scheduling has to do with the FM Team and how they are spending their work days. For instance, the FM Team should begin their involvement with the FM process by breaking their current connections with day-to-day activities. If they fail to take this step, it will be difficult for the team to transition into their new roles.

As the FM process is first introduced and then matures, the

FM team's roles will develop, as will the ways in which they spend their time. At first, they will be primarily involved in creating maps and investigating failures that occurred some time in the past. At this point, the team members may have relatively little to do with current day-to-day activities. As the FM process matures and the FM database becomes more complete and usable, the team will play a role in day-to-day maintenance that will become increasingly intertwined with current problems.

The first person to deal with Malfunction Reports will soon be the Diagnostician, and the second person will be the Troubleshooter. Their involvement will ensure that all repairs are completed quickly and accurately, and that key information is captured for future use.

By the time the FM process has completely matured, the FM team will have become an integral part of the Maintenance Team.

Where to Start – Where to Proceed

The last section of this chapter deals with providing directions: Where to Start and Where to Proceed from there. Teams made up of strong performers will ultimately find the right direction in almost any pursuit and begin making progress. The problem is that new programs often die waiting for that sequence to occur. It is better to provide direction early on and allow the team to get their bearings while making progress.

The starting point for implementing the FM process is mapping historic failures. Many of the current generation of Computer Maintenance Management Systems (CMMS) already provide for retaining the proper information needed to identify Malfunction Reports and Failure Modes. Some systems call these Initial Failure Codes and Final Failure Codes. Even in situations where the assigned codes have no true basis in fact or reality, there is typically other information that can be useful in determining the appropriate Malfunction Report and Failure Mode. For instance, the narrative description of any failure that is found in many CMMS's, contains enough of a description to determine the correct Malfunction Report. Also, the material and parts consumption records from either the CMMS or warehouse records should be helpful in identifying the component used to close the specific failure record.

Involving the entire FM Team in assembling the first failure maps will ensure that the entire team will become familiar with both the system and the discipline being applied.

After the first failure maps have been developed, the next step should be for the Failure Analyst and Root Cause Analyst to begin investigating some recent repairs. This work will serve several purposes:

1. It will provide data to finalize a number of failure maps.
2. It will serve to give everyone a "dose of reality" in terms of how well the current maintenance process is actually working.
3. It will serve notice to the remainder of the organization about the level of follow-up that is coming soon.

As the database becomes more populated and the FM Team becomes more familiar with how the maintenance process currently works and how it will change under a FM environment, the FM team should begin taking over handling a small portion of each day's failures. As they become more proficient and increasingly comfortable in their roles, the team should add more and more of each day's failures to their workload until finally, all failures are being handled using the FM process.

Chapter 8

Conclusion

To exist is to change, to change is to mature, to mature is to go on creating oneself endlessly. ***Henri Bergson***

This book started by focusing on how to assess the value that Failure Mapping can provide to your operation and ended with how to go about implementing a Failure Mapping process. For many readers, this book may be their first detailed exposure to the concept of Failure Mapping. Although the individual elements of the process are not new, there are several characteristics of the FM process that are new. These new aspects include:

1. Wrapping all the elements into a single cohesive and comprehensive process.
2. "Tightening" up the definitions so that the Malfunction Report and the Failure Mode can be used as starting points and ending points of failure maps.
3. Recognition that failure maps are consistent, recurring patterns to which statistics can be applied.
4. Recognition that failure maps are tools that can be used to make reactive maintenance much more efficient.
5. Recognition that failure maps, when followed through to the logical conclusion, can be used as tools to improve system reliability by identifying Failure Mechanisms and Root Causes.
6. Recognition that Failure Mapping can be an effective tool

139

in forcing the transition from a reactive environment to a proactive environment.

As with many of the business process tools that have been introduced into the workplace in the last several years, Failure Mapping is a system that is based on increased structure and discipline. One element of structure that is critical to the effectiveness of the FM process is the terms used to describe key elements like Malfunction Report and Failure Mode. Equally important with the design structure of these terms is the discipline used during their application.

For instance, weeks may be spent in determining precisely which terms to use, then having the database contaminated by users who feel free to use more "colorful" terms during an application.

A significant part of the Return on Investment report that is available from applying the FM process is obtained simply by applying the individual parts of the process. For instance, if the system is used to investigate and determine the Failure Mechanism associated with a portion of the failures, it is possible to make some improvement in reliability by eliminating them. On the other hand, there are many elements of the FM process that are synergistic. In other words, the whole is greater than the sum of the individual parts. Without implementing the complete FM process, it will be impossible to harvest all the benefits.

Before closing, there is one area where criticism of the ideas expressed in this book is likely. That is the expectation that "all" failures can be exposed to Failure Analysis and Root Cause Analysis. Some people will say that it is not cost effective or not possible to perform Failure Analysis or Root Cause Analysis for a specific volume of failures. Their point is that it simply takes too much time and resources.

People who believe that Failure Analysis and Root Cause Analysis consume large amounts of time and resources are victims of a self-fulfilling prophesy. The cumbersome nature of their approach creates a situation where very few failures are investigated. When very few events are investigated, individuals remain suspicious and unwilling participants in the work. If it is decided that things should be done more quickly and efficiently and a "federal case" should not be made out of every finding, the system can act much more efficiently and effectively. Take for instance, situations where blame is not a part of the paradigm. People quickly accept their role as a cause, say "my bad" and move on. If the environment is one where it is necessary to pry evidence out of people, the message is that the people expect their evidence will be misused.

The FM process is designed to handle all failures, that is, a large volume of problems on a real-time basis. Success depends on participants being candid, and leaders acting in a mature and good-natured manner.

Appendix
Failure Mapping -
A Quick Description

> **Failure Mapping (FM)**
>
> A description of all the steps that lead up to the failure of a physical system and the analysis needed to recover from it. FM uses a structure that includes well-defined elements for each specific step in the failure-recovery process.

How does FM work?

• FM recognizes the extended cause-effect relationship that exists whenever a system failure occurs.

• FM recognizes that within the extended cause-effect relationship, there are two highly-recognizable elements that are most visible when dealing with any system failure. The first is the Malfunction Report. The Malfunction Report describes the system function that is no longer being performed adequately, and the specific behavior that is being portrayed. The second is the Failure Mode. The Failure Mode describes the specific component that has developed a failure-causing defect and the specific condition of the component that encompasses the defect.

• FM recognizes that there might be more than one Failure Mode (component-condition combination) associated with each Malfunction Report (function-behavior). Over time, the number of times that each distinct Failure Mode occurs can be recorded, and those records can be used to calculate the relative likelihood of each.

• FM also recognizes that physical systems, particularly those with sophisticated computer control, have the ability to provide clues as to which of the possible failure modes are most likely to have produced the current Malfunction Report.

• FM recognizes that a well-organized and comprehensive set of failure maps is highly valuable in both responding to failures and taking the steps needed to prevent failures.

Why does FM work?

• The first reason why FM is effective is because there are no new Failure Modes or Failure Mechanisms. What has happened is likely to happen again.

• In addition, the science of cause and effect and deterioration leading to a failure-causing defect is not new. It is based on consistent, well-established patterns.

• The patterns leading to failures are made up of elements that, once clearly defined, can be observed, recorded and analyzed.

How is Value derived from FM?

• If it is possible to identify the possible Failure Modes as soon as a Malfunction Report is received, it might be possible to make simple repairs (e.g. resetting a computer) that can restore the affected function.

• If more than one level of repair is possible, knowing the most likely eventual repair will aid in selection of the proper level at the start.

• Knowledge of the most likely cause of a breakdown can reduce the time required for troubleshooting.

• Knowledge of the most likely eventual repair can reduce the cost and the amount of resources spent on that repair.

• As work management processes become increasingly better organized, the value of knowing the likelihood of an eventual outcome will continue to grow. (In other words, if the work management system is designed to capture the benefits associated with the ability to plan and schedule job A, then there is a significant value in being able to dispatch crews to perform job A, rather than to determine if job A, job B, job C or job D needs to be done.)

How is FM used?

• FM begins with creating an organization that functions to collect and use the information contained within individual failure maps, and the overall FM system. Five key functions accomplished by this organization are:

  1. **Failure Mapping** – The individual performing this function analyzes files showing the Malfunction Reports and the Failure Modes used to close out those reports. He utilizes his expertise and experience to add data needed to populate the other FM elements.

  2. **Diagnostics** – The individual performing diagnostics provides a "back office" function that uses the FM database and additional data collected remotely to identify which Failure Modes are possible, what additional infor-

mation is needed to determine the most likely Failure Mode, and the best path forward. He performs the function of dispatching resources and providing instructions concerning the steps that should be taken and the best sequence for them to be used.

3. Troubleshooting – This diagnostician also uses the information to conduct the search for the defect. Troubleshooting is often a costly, invasive procedure, so the sequence is critical. Typically, the most likely candidate is pursued first, but occasionally, low-cost and quickly-completed options that have a lower likelihood are completed first, just to eliminate the possibility of an "easy fix". The end product of effective troubleshooting is identification of the component containing the defect causing the failure, and the specific condition that constitutes the defect. This work is not complete until the Failure Mode (component and condition) are recorded in the FM system.

4. Failure Analysis – Each and every failure is caused by one of the possible Failure Mechanisms. For example, mechanical devices fail by one of only four Failure Mechanisms (corrosion, erosion, fatigue, or overload). The person performing Failure Analysis identifies the specific Failure Mechanism that ultimately resulted in the defect that caused each failure.

5. Root Cause Analysis – Each and every failure has at least one cause at each of three levels. To permanently eliminate the cause of failures it is important that Root Cause Analysis be conducted and corrective action taken. The following are the three levels of cause:

> • **Physical Cause** – Physical cause is the problem that allows the Failure Mechanism to become active. (For example, if the failure mechanism is fatigue, a source of repetitive tension-compression cycles must

have existed. Examples can be imbalance or mis-alignment.)

• **Human Cause** – The human cause is the specific individual whose action or inaction resulted in the presence of the physical cause. (Although this individual is not the human cause, this does not imply that he is at fault unless he intentionally disregarded established procedures or practices.)

• **Latent or Systemic Cause** – The Latent or Systemic cause is the "trap" set by the current system or organization that allowed the individual to fail. The trap may be inadequate procedures, acceptance of poor practices, or any other circumstance that provides an environment in which actions or inactions leading to failure are tolerated or encouraged.

The specific organization used to apply FM can vary, but all the functions described above need to be accomplished. Some organizational arrangements and staffing choices are more conducive to successful application of FM. For instance, "doers" are seldom effective "investigators". (For example, individuals charged with performing repairs are seldom effective in performing failure analysis or root cause analysis because the results from those activities are not tangible and "doers" are people who value tangible results.)

How to construct a Failure Map

• Construction of a failure map begins with establishing the system in which failure mapping data will be entered. The FM system includes the basic structure of the failure map being used and the definition of each of its elements.

The following is an example of a FM structure.

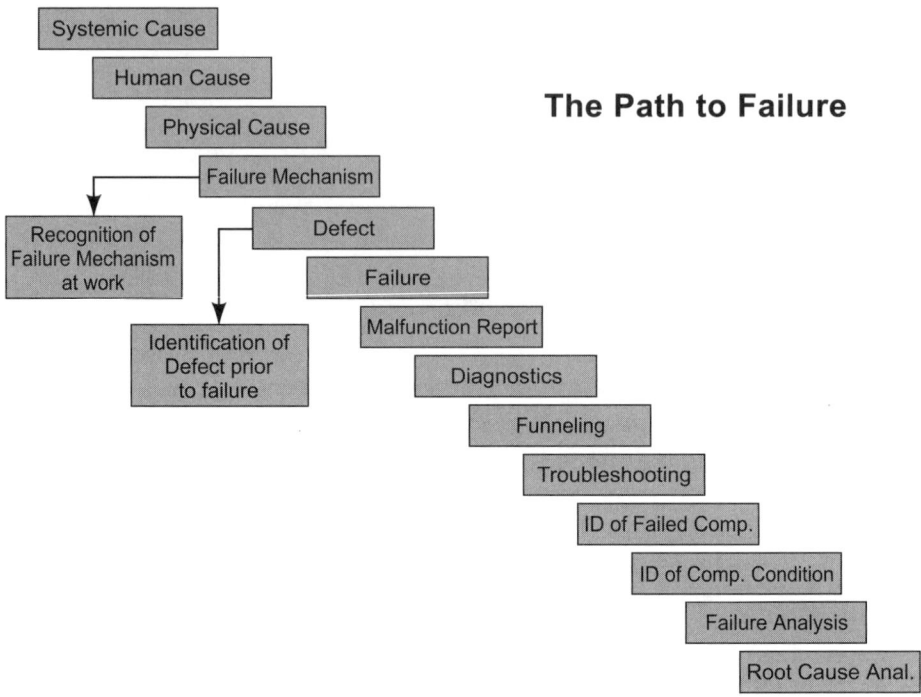

The Path to Failure

• The following are usable definitions for each element.

  • Systemic Cause – The weakness or "trap" in the systems or organization that causes or allows the individual to act in a manner that creates a physical cause.
  • Human Cause – The specific action or inaction taken by an individual that result in the physical cause.
  • Physical Cause – The specific physical condition that allows a Failure mechanism to begin causing deterioration.
  • Failure Mechanism – The specific deterioration-causing mechanism that ultimately results in a failure-causing defect.
  • Recognizing Failure Mechanisms at Work – Failure Mechanisms provide evidence of their existence. If rec-

ognized in time, the Failure Mechanism can be halted and the failure prevented.

• Defect – After a Failure Mechanism has been at work for an adequate period, a defect will form. Under the appropriate conditions, the defect will result in a failure.

• Recognizing Defects Before Failure – For many failures there is a time delay between formation of the defect and occurrence of the failure. During this time delay, it is possible to identify and eliminate the defect, thus preventing the failure.

• Failure – The failure is the specific event when the system being studied can no longer perform one or more of its required functions.

• Malfunction Report – After the failure, some individual reports what has happened. To be useful, the malfunction report must be in a usable format. The format must be accurate, descriptive, and in a form that can be processed. The combination of Function and Behavior fits those standards.

• Diagnostics – In a mature FM-based system, diagnostics begins with looking at the FM database to determine which Failure Modes are possible, based on the reported Malfunction Report. The diagnostician uses the FM database to help determine the best strategy. Depending on the information contained in the database, several directions are possible. The diagnostician should pick the one leading to the quickest repair and the least costs.

• Funneling - If several failure modes are possible, the diagnostician will determine if some forms of additional information will help in reducing the possible alternatives. He will also understand the sources of that information. A call to the equipment operator may provide a description of additional symptoms. A review of maintenance history might show recent failures or chronic failure modes. The objective of funneling is to set a direction for troubleshooting likely to deliver the most accurate and effective results.

• Troubleshooting – Troubleshooting is a hands-on function that provides an opportunity for a person to identify a defect. This procedure is typically invasive, so it takes time and money to complete.

• Identification of a Failed Component – The first half of identifying the Failure Mode is identifying the defective component.

• Identification of Component Condition Resulting in Failure – This step is the second half of identifying the Failure Mode, which is a description of the specific condition associated with the defect. (e.g. overheated). Occasionally parts are exchanged and the function restored without determining which component is defective. The component name should then be listed, along with the phrase: "No Identified Defect – Function Restored When Component was Exchanged" (NID).

• Failure Analysis – Failure analysis is the step of identifying the Failure Mechanism that resulted in the failure-causing defect. For example, corrosion or fatigue. The Physical Cause usually links directly to the Failure Mechanism, so it is important to take the time to read the clues that help identify the Failure mechanism.

• Root Cause Analysis – Root Cause Analysis provides three levels of cause: Physical, Human, and Systemic. These causes are described above, and performing the investigation as a part of failure close-out is the only way that these causes can be identified and corrected.

Recognizing that there is a "one-to-many" relationship between Malfunction Reports and Failure Modes, it is best to begin assembling data, using a MS Access database that will facilitate adding multiple files to each function over an extended period of time.

What does a Failure Mapping System look like?

A typical input form used to collect data on failures or

Malfunction Reports, and to convert the data into usable failure maps looks like the example below:

Failure Reporting and Mapping Input Form

Category	Item	Sub-item
Initial Report	Date of Event	
	Equipment Identification Number	
	Narrative Description of Event	
Malfunction Report	Affected Function	
	Behavior	
	Funneling Data	Physical Symptoms
		Earlier Behavior
		Computer Alerts
Mapping Linkages	Systems Where Failures can Cause this Malfunction Report	
	Subsystems or Components Where Failure can Cause this Malfunction Report	
Failure Mode	Failed Component	
	Component Condition	
Failure Analysis	Failure Mechanism	Mechanical
		Electrical
Root Cause Analysis	Physical Cause	
	Human Cause	
	Systemic Cause	

A typical FM database that can be used by the diagnostician to quickly access needed information is shown below. (The data is portrayed in the form of a spreadsheet. It would be even more convenient to provide the information in the form of an automated database that would provide realistic alternatives when the person performing the diagnostics started to enter data.)

Although this report is shown in the form that the information might take if it was being reported via an Access database, it is equally possible to provide the information in spreadsheet form. It is envisioned that the information from the database will be generated as additional fields are completed. For instance, adding an equipment number might limit the information being provided to whatever is appropriate to the specific fleet or system to which that equipment item belongs.

Failure Mapping Report Form		
Equipment Number		
Fleet to Which this Item Belongs		
Malfunction Report	Function Provided by this System	
	Possible Behaviors during Loss of Function	
Funneling Support	Systems where a failure can produce this behavior	
	Symptoms, data or alerts that point to a specific system	
	Subsystems or Components that can produce this behavior	
	Symptoms, data or alerts that point to a specific system	
Most Likely Failure Mode	Most Likely Component Causing This Event	
	Most Likely Component Condition	
Alternatives	Alternatives and Order of Likelihood	

Obviously, different formats and different data structures are most appropriate with different situations. It is left to the reader to select the information and structure that best suits the purpose.

Helpful hints

• The failure mapping process should start by identifying the few functions being performed by the system being mapped. Care should be taken not to identify functions that are not critical to the operation or the products being provided by that system. Do not identify functions that in some way duplicate or overlap other functions.

• For each function it is important to describe the several ways it can "misbehave". When making the list of behaviors, set up a single consistent list that can be used uniformly across all functions. Avoid including behaviors that are similar or duplicates of others that are already in the list. Choose the single behavior that best describes the widest range of behaviors. Using more than one term to represent the same behavior will dilute the count of those failures that are the same, and thus provide an inaccurate portrayal of failure statistics.

• The same problem with use of more than one term exists with components and component conditions. One and only one term should be used to describe a single noun or a single adjective. Using redundant terms that have the same or similar meanings will dilute statistics and provide improper directions as to how situations should be addressed. An example of a condition that is frequently described using different terms is "overheated". The alternative terms include overheated, burned, burnt, burned up, charred, cooked, fried, and a variety of others. All these words have a meaning suggesting that the component failed as the result of experiencing heat or temperature beyond its capabilities.

• When a specific set of terms to be used has been selected, they should be placed in look-up boxes in the database and the selection process confined to only the ones in the table.

• After the FM system has been set up, it will be necessary to populate the database. A simple starting point is to record the elements of the failure maps that are already known.

Most Computer Maintenance Management Systems (CMMS) should have records of each failure that contain (or can be interpreted to provide) a historic record of Malfunction Reports (affected function and behavior when failed) and the Failure Modes (failed component and condition of the failed component) that were used to restore the function and close out the repair. Once an individual file is initiated for each instance, it will be the job of the individual who constructs the failure maps to "fill in the gaps" by adding the other information needed to aid future diagnosis and troubleshooting.

References for Further Reading:

Abernethy, Dr. Robert B.; *The New Weibull Handbook – Fifth Edition;* Robert B. Abernethy, North Palm Beach, Florida; 2004

Block, Peter; *Flawless Consulting*; Pfeiffer & Company, San Diego CA, 1981

Conner, Daryl R., *Managing at the Speed of Change,* Villard Books, New York, 1992

Daley, Daniel T.; *The Little Black Book of Reliability Management;* Industrial Press, New York; 2007

Daley, Daniel T.; *The Little Black Book of Maintenance Excellence;* Industrial Press, New York; 2008

Daley, Daniel T.; *Understanding the Path to Failure and Benefitting from that Knowledge;* SKF Reliability Systems @ptitude Exchange Article, http://www.aptitude exchange.com, February 2008

Hammer, Michael; *The Process Audit*; Harvard Business Review Article; April 2007

Gillovich, Thomas; *How We Know What Isn't So; Free Press, 1993*

Ireson, W. Grant & Coombs, Clyde F. & Moss, Richard Y. ; *Handbook of Reliability Engineering and Management – Second Edition;* McGraw-Hill, New York; 1996

Latino, Robert J. & Latino, Kenneth C. ; *Root Cause Analysis – Improving Performance for Bottom Line Results;* CRC Press, New York; 1999

155

Markova, Dawna, Ph.D.; The Open Mind – Exploring the Six Patterns of Natural Intelligence; Conari Press; Berkeley, CA; 1996

Nyman, Don and Levitt, Joel; *Maintenance Planning, Scheduling and Coordination;* Industrial Press, New York; 2001

O'Connor, Patrick D. T. ; *Practical Reliability Engineering – Fourth Edition;* John Wiley & Sons, LTD; West Sussex, England; 2002

Senge, Peter M.; *The Fifth Discipline, The Art & Practice of the Learning Organization;* Doubleday Publishing; New York; 1990

Wireman, Terry; *Computerized Maintenance Management Systems*; Industrial Press, New York; 1994

Wireman, Terry; *Inspection and Training for TPM*; Industrial Press, New York; 1992

Wireman, Terry; *World Class Maintenance Management*; Industrial Press, New York; 1990

Index